Additive Manufacturing

The text explores the development, use, and effect of additive manufacturing and digital manufacturing technologies for diverse applications. It will serve as an ideal reference text for graduate students and academic researchers in diverse engineering fields including industrial, manufacturing, and materials science.

This book:

- Discusses the application of 3D virtual models to lasers, electron beams, and computer-controlled additive manufacturing machines.
- Covers applications of additive manufacturing in diverse areas including healthcare, electronics engineering, and production engineering.
- Explains the use of additive manufacturing for biocomposites and functionally graded materials.
- Highlights rapid manufacturing of metallic components using 3D printing.
- Illustrates production and optimization of dental crowns using additive manufacturing.

This book covers recent developments in manufacturing technology, such as additive manufacturing, 3D printing, rapid prototyping, production process operations, and manufacturing sustainability. The text further emphasizes the use of additive manufacturing for biocomposites and functionally graded materials. It will serve as an ideal reference text for graduate students and academic researchers in the fields of industrial engineering, manufacturing engineering, automotive engineering, aerospace engineering, and materials science.

Mathematical Engineering, Manufacturing, and Management Sciences

Series Editor:
Mangey Ram, Professor, Assistant Dean (International Affairs), Department of Mathematics, Graphic Era University, Dehradun, India

The aim of this new book series is to publish the research studies and articles that bring up the latest development and research applied to mathematics and its applications in the manufacturing and management sciences areas. Mathematical tool and techniques are the strength of engineering sciences. They form the common foundation of all novel disciplines as engineering evolves and develops. The series will include a comprehensive range of applied mathematics and its application in engineering areas such as optimization techniques, mathematical modelling and simulation, stochastic processes and systems engineering, safety-critical system performance, system safety, system security, high assurance software architecture and design, mathematical modelling in environmental safety sciences, finite element methods, differential equations, reliability engineering, etc.

Advanced Manufacturing Processes
Edited by Yashvir Singh, Nishant K. Singh, and Mangey Ram

Additive Manufacturing
Advanced Materials and Design Techniques
Pulak Mohan Pandey, Nishant K. Singh, and Yashvir Singh

Advances in Mathematical and Computational Modeling of Engineering Systems
Mukesh Kumar Awasthi, Maitri Verma, and Mangey Ram

Biowaste and Biomass in Biofuel Applications
Edited by Yashvir Singh, Vladimir Strezov, and Prateek Negi

For more information about this series, please visit: www.routledge.com/Mathematical-Engineering-Manufacturing-and-Management-Sciences/book-series/CRCMEMMS

Additive Manufacturing
Advanced Materials and Design Techniques

Edited by
Pulak Mohan Pandey
Nishant K. Singh
and Yashvir Singh

CRC Press is an imprint of the
Taylor & Francis Group, an **informa** business

First edition published 2023
by CRC Press
6000 Broken Sound Parkway NW, Suite 300, Boca Raton, FL 33487-2742

and by CRC Press
4 Park Square, Milton Park, Abingdon, Oxon, OX14 4RN

CRC Press is an imprint of Taylor & Francis Group, LLC

© 2023 selection and editorial matter, Pulak Mohan Pandey, Nishant K. Singh and Yashvir Singh; individual chapters, the contributors

Reasonable efforts have been made to publish reliable data and information, but the author and publisher cannot assume responsibility for the validity of all materials or the consequences of their use. The authors and publishers have attempted to trace the copyright holders of all material reproduced in this publication and apologize to copyright holders if permission to publish in this form has not been obtained. If any copyright material has not been acknowledged, please write and let us know so we may rectify in any future reprint.

Except as permitted under U.S. Copyright Law, no part of this book may be reprinted, reproduced, transmitted, or utilized in any form by any electronic, mechanical, or other means, now known or hereafter invented, including photocopying, microfilming, and recording, or in any information storage or retrieval system, without written permission from the publishers.

For permission to photocopy or use material electronically from this work, access www. copyright.com or contact the Copyright Clearance Center, Inc. (CCC), 222 Rosewood Drive, Danvers, MA 01923, 978-750-8400. For works that are not available on CCC, please contact mpkbookspermissions@tandf.co.uk

Trademark notice: Product or corporate names may be trademarks or registered trademarks and are used only for identification and explanation without intent to infringe.

ISBN: 9781032192635 (hbk)
ISBN: 9781032192659 (pbk)
ISBN: 9781003258391 (ebk)

DOI: 10.1201/9781003258391

Typeset in Sabon
by codeMantra

Contents

Preface		vii
Editors		ix
Contributors		xi

1 **Effect of process parameters on mechanical properties of additively manufactured metallic systems**   1
LAKHINDRA MARANDI AND INDRANI SEN

2 **Parametric study of fused deposition modelling**   21
KRITI SRIVASTAVA AND YOGESH KUMAR

3 **Microstructural and mechanical properties of aluminium metal matrix composites developed by additive manufacturing—A review**   43
AMARISH KUMAR SHUKLA AND J. DUTTA MAJUMDAR

4 **In situ process monitoring and control in metal additive manufacturing**   57
MUKESH CHANDRA, VIMAL K.E.K., AND SONU RAJAK

5 **Additive manufacturing: Materials, technologies, and applications**   77
RAJ AGARWAL, SHRUTIKA SHARMA, VISHAL GUPTA, JASKARAN SINGH, AND KANWALJIT SINGH KHAS

6 **A case study on the role of additive manufacturing in dentistry**   99
RAHUL JAIN, SUDHIR KUMAR SINGH, AND RAJEEV KUMAR UPADHYAY

vi Contents

7 Role of additive manufacturing in biomedical application 117

VIPIN GOYAL, AJIT KUMAR, AND GIRISH CHANDRA VERMA

8 Wire arc additive manufacturing of titanium alloys:
A review on properties, challenges, and applications 133

SUJEET KUMAR AND VIMAL K.E.K.

9 Advances in additive manufacturing 149

M.B. KIRAN AND V.J. BADHEKA

10 Additive manufacturing of polymer-based functionally
graded materials 167

MOHIT KUMAR, NEHA CHOUDHARY, AND VARUN SHARMA

11 Processing techniques, principles, and applications of
additive manufacturing 187

SUMIT KUMAR SHARMA, RANJAN MANDAL, AND
AMARISH KUMAR SHUKLA

Index 203

Preface

Advanced additive manufacturing techniques put forth technological developments in materials science and conventional manufacturing processes. The fast-growing additive/3D printing technology and the industrial Internet of Things aim to create practical solutions for industry. This book explores the development, use, and effect of additive manufacturing and digital manufacturing technologies. It raises technological and business-oriented developments with others in product design and design studies. It contains a combination of functional and transdisciplinary patterns and is enriched in case and design content. The chapters cover a number of design-based perspectives on additive manufacturing that are seldom discussed in major conferences and journals, which are still mainly and most significantly associated with tools, innovations, and technological progress. This book incorporates new, unpublished, and very well-written relevant research articles and studies describing the latest findings and progress in the field of "Advanced Additive Manufacturing Processes" and has a tremendous relevance in the lives of academic institutions, practitioners, researchers, and industry players.

Editors

Dr. Pulak Mohan Pandey completed his PhD in Additive Manufacturing/3D Printing from IIT Kanpur in 2003. He joined IIT Delhi as a faculty member in 2004 and is presently a professor. In IIT Delhi, Dr. Pandey diversified his research areas in micro- and nano-finishing and micro-deposition and continued working in the area of 3D printing. He supervised 35 PhDs and more than 36 MTech theses in the last 10 years and filed 21 Indian patent applications. To his credit, he has published approximately 185 international journal papers and 45 international/national refereed conference papers.

Dr. Nishant K. Singh is an associate professor at Harcourt Butler Technical University, Kanpur, Uttar Pradesh, India. He earned his PhD from IIT, Dhanbad, and Master's in Production Engineering and BTech in Mechanical Engineering from Delhi College of Engineering. Dr. Singh has more than 16 years of teaching experience. He has written more than 90 research articles in well-known international journals and serves as a reviewer and editorial member in peer-reviewed journals and conferences. His research interests include tribology, micro-manufacturing, non-conventional machining, and additive manufacturing processes.

Dr. Yashvir Singh is presently a postdoctoral fellow at Universiti Tun Hussein Onn Malaysia, Parit Raja, Batu Pahat, Johor, Malaysia. He is also an associate professor in the Department of Mechanical Engineering, Graphic Era Deemed to be University, Dehradun, Uttarakhand, India. Dr. Singh completed his PhD at the University of Petroleum and Energy Studies, Dehradun, Uttarakhand, India. He has more than 15 years of teaching experience. Dr. Singh has written more than 120 research articles and published them in various peer-reviewed journals. He is also a reviewer and editorial board member in various journals. Dr. Singh works in tribology, bioenergy, lubrication, and manufacturing.

Contributors

Raj Agarwal
Department of Mechanical
  Engineering
Thapar Institute of Engineering
  and TechnologyPatiala, India

V.J. Badheka
Department of Mechanical
  Engineering
Pandit Deendayal Energy University
Gandhinagar, India

Mukesh Chandra
BIT Sindri
Jharkhand, India

Neha Choudhary
Additive and Subtractive
  Manufacturing Lab
Indian Institute of Technology
Roorkee, India

J. Dutta Majumdar
Department of Metallurgical &
  Materials Engineering
Indian Institute of Technology
Kharagpur, India

Vipin Goyal
Department of Mechanical
  Engineering
Indian Institute of Technology
Indore, India

Vishal Gupta
Department of Mechanical
  Engineering
Thapar Institute of Engineering
  and Technology
Patiala, India

Rahul Jain
Department of Mechanical
  Engineering
Galgotias University
Gautam Buddha Nagar, India

Kanwaljit Singh Khas
Product and Industrial Design
  Department
Lovely Professional University
Punjab, India

M.B. Kiran
Department of Mechanical
  Engineering
Pandit Deendayal Energy University
Gandhinagar, India

xii Contributors

Ajit Kumar
Department of Mechanical
    Engineering
Indian Institute of Technology
Indore, India

Mohit Kumar
Additive and Subtractive
    Manufacturing Lab
Indian Institute of Technology
Roorkee, India

Sujeet Kumar
Department of Mechanical
    Engineering
National Institute of Technology
Patna, India

Yogesh Kumar
Department of Mechanical
    Engineering
National Institute of Technology
Patna, India

Ranjan Mandal
Department of Mechanical
    Engineering
BIT Sindri
Jharkhand, India

Lakhindra Marandi
Department of Metallurgical
    Engineering
Indian Institute of Technology
    (BHU)
Varanasi, India

Sonu Rajak
Department of Mechanical
    Engineering
National Institute of Technology
Patna, India
and
BIT Sindri
Jharkhand, India

Indrani Sen
Department of Metallurgical and
    Materials Engineering
Indian Institute of Technology
Kharagpur, India

Shrutika Sharma
Department of Mechanical
    Engineering
Thapar Institute of Engineering
    and Technology
Patiala, India

Sumit Kumar Sharma
Department of Mechanical
    Engineering
BIT Sindri
Jharkhand, India

Varun Sharma
Department of Mechanical
    Engineering
Indian Institute of Technology
Roorkee, India

Amarish Kumar Shukla
Deptment of Metallurgical &
    Materials Engineering
Indian Institute of Technology
Kharagpur, India

Jaskaran Singh
Department of Mechanical
    Engineering
Thapar Institute of Engineering
    and Technology
Patiala, India

Sudhir Kumar Singh
Department of Mechanical
    Engineering
Galgotias University
Gautam Buddha Nagar, India

Kriti Srivastava
Department of Mechanical
  Engineering
National Institute of Technology
Patna, India

Rajeev Kumar Upadhyay
Department of Mechanical
  Engineering
Hindustan College of Science &
  Technology
Mathura, India

Girish Chandra Verma
Department of Mechanical
  Engineering
Indian Institute of Technology
Indore, India

Vimal K.E.K.
Department of Production
  Engineering
National Institute of Technology
Tiruchirappalli, India

Chapter 1

Effect of process parameters on mechanical properties of additively manufactured metallic systems

Lakhindra Marandi
Indian Institute of Technology (BHU)

Indrani Sen
Indian Institute of Technology (Kharagpur)

CONTENTS

List of abbreviations and symbols	1
1.1 Metal additive manufacturing techniques	2
1.2 Influence of additive manufacturing process parameters	4
1.3 Effect of processing parameters on mechanical properties	5
1.3.1 Laser power (P) and scan speed (s)	6
1.3.2 Hatch spacing (h) and layer thickness (t)	7
1.3.3 Scan orientation (θ) and test orientation (θ')	9
1.3.4 Deposition depth (H)	10
1.4 Case study	11
1.5 Limitation and future scope	14
1.6 Closure	14
References	14

LIST OF ABBREVIATIONS AND SYMBOLS

AM	Additive manufacturing
CAD	Computer automated design
d	Laser diameter
DMLS	Direct metal laser sintering
EBDM	Electron beam direct manufacturing
EBM	Electron beam melting
ED	Laser energy density
EL	Linear energy density
h	Hatch spacing
H	Deposition depth
LC	Laser consolidation

DOI: 10.1201/9781003258391-1

2 Additive Manufacturing

LD   Laser deposition
LENS   Laser engineered net shaping
P   Laser power
PTAS   FFF plasma transferred arc selected free form fabrication
s   Scan speed
SHS   Selective heat sintering
SLS   Selective laser sintering
t   Layer thickness
θ Scan orientation
θ' Test orientation

1.1 METAL ADDITIVE MANUFACTURING TECHNIQUES

Metallic materials used for structural or functional purposes are manufactured through conventional or advanced processing routes. Conventional processes are used for bulk production with better control on composition and microstructure. On the other hand, advanced materials processing is used for producing complex components. Additive manufacturing (AM) is a state-of-the-art technique that is capable of producing components with intricate shapes, compositional variation, and required porosities. It is an emerging technology that has been developed to manufacture near net shape products with cost effectiveness and the desired properties [1–5]. AM processes for metallic materials use a high-power energy source (laser beam, electron beam, etc.) to melt the metal powder and thereby to deposit it in a layer-by-layer fashion in an inert chamber. In fact, AM is one of the advanced techniques which offers a wide range of microstructural variation and corresponding mechanical properties. Altering the process parameters further provides the scope to vary the mechanical properties of the manufactured materials as well. This chapter highlights the effect of process parameters on the mechanical properties of the component produced using AM techniques.

Several types of AM techniques are available for fabricating metallic systems. On the basis of their mode of operation, these can be categorized as follows: (i) powder bed system, (ii) powder feed system, and (iii) wire feed system.

(i) **Powder bed system:** Powder bed system is an AM technique where a high-power laser beam is used to melt pre-alloyed metal powder spread over a platform. The high-power energy source is incident on the first layer of powder and it melts the metal powder according to the computer automated design. After the first layer of the deposition or sintering, successive layer is spread over the first layer and excessive materials are skimmed off the platform. After that, the laser deposits the second layer and the process repeats itself. Powder bed system is further subdivided based on energy sources into thermal fused selective heat sintering (SHS), laser beam–based direct metal laser sintering (DMLS), selective laser sintering (SLS), and electron beam sourced electron beam melting (EBM) [2,4,6,7]. However, relatively longer print time and post-processing treatments are some of the drawbacks of the powder bed AM methods.

(ii) **Powder feed system:** Among all the different AM techniques, powder feed is the most widely used and convenient method [8–11]. According to this method, high-power laser beam and pre-alloyed metal powder are simultaneously injected and sprayed upon the substrate using multiple nozzles. The setup is schematically shown in Figure 1.1. The injected powder melts initially and subsequently solidifies as the laser moves away. The powder feed system works on a movable platform with stationary laser or in the vice versa mode. The mobile platform usually can move in three axes or five axes, which helps in making any intricate design. The advantages of the powder feed system are faster speed, low cost of production, lesser powder wastage, and the ability to rebuild the damaged parts from the working components. Few examples of powder feed systems are laser engineered net shaping (LENS), direct metal deposition (DMD), laser consolidation (LC), and laser deposition (LD) [12–16].

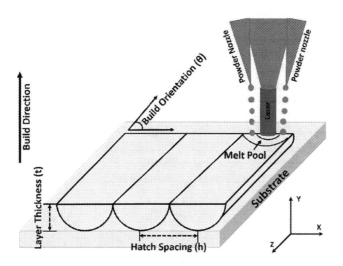

Figure 1.1 Schematic illustrating the AM method (powder feed AM).

4 Additive Manufacturing

(iii) **Wire feed system:** In the wire feed system, instead of powder, a solid wire acts as the input material. An energy source such as the laser is used to melt the wire which solidifies as the laser moves away. The process keeps on repeating until a three-dimensional structure is formed. The wire feed system is further subdivided into three categories based on the types of energy sources, namely, laser beam, electron beam, and welding type [17]. Some of the famous wire feed systems are electron beam direct manufacturing (EBDM), plasma transferred arc selected free form fabrication plasma transferred arc selected free form fabrication, etc. However, this process lacks in inducing porosities as can be achieved by other methods [14,18,19].

1.2 INFLUENCE OF ADDITIVE MANUFACTURING PROCESS PARAMETERS

It is important to mention that despite different additive technologies being involved in the manufacturing of metallic materials, the process parameters for almost all the cases are nearly the same. In fact, all the laser-based AM techniques have some common mode of operations and parameters. Nevertheless, based on the operation, equipment, and technologies, different materials can be manufactured through AM. In this section, the effect of AM process parameters on the mechanical properties of two different categories of metallic systems is considered. These are (i) the most widely used structural material, i.e., stainless steel, and (ii) a specialized functional material, NiTi-based shape memory alloy.

It is noteworthy that stainless steel is the most common material that is extensively used for structural purposes. The AM technique itself, as well as any variation in the manufacturing process parameters, substantially influences the microstructural evolution in the material. This is reflected in obtaining variation in the related mechanical properties of stainless steel as well. On the other hand, NiTi is a smart material that is capable of remembering its shape under the influence of temperature and/or stress. A solid-state phase transformation from parent austenite phase to product martensite results in this shape recovery. This unique material, in fact, is used for both its structural and functional properties. Alterations in the AM process parameters influence microstructural and compositional variations of NiTi alloy. This, in turn, leads to significant changes in the transformation behavior and consequently the mechanical properties of the material [20,21].

The present chapter highlights the role of different process parameters on the mechanical properties of additively manufactured NiTi and stainless steel alloys. An extensive literature review is pursued in this regard to generate a holistic insight about the possible conditions that can alter the performance of these materials. The pros and cons of varying the individual process parameters in attaining improved as well as uniform characteristics are documented in this chapter. Nevertheless, any discrepancy in the trend is also highlighted.

1.3 EFFECT OF PROCESSING PARAMETERS ON MECHANICAL PROPERTIES

Processing parameters play dominating roles in modifying the mechanical properties of additively manufactured materials. Mechanical properties of the AM processed materials primarily depend on the composition, quality, and morphology of the feed materials (powder). Mechanical properties of the materials also vary with altering different processing parameters such as laser power (P), scan speed (s), laser diameter (d), hatch spacing (h), layer thickness (t), scan orientation (θ), test orientation (θ'), and deposition depth (H). The entire response is optimized by taking into account all of the essential parameters in terms of laser energy density (E_D) and is provided using equation (1.1) [12].

$$E_D = \frac{P}{s \times d} \quad J/mm^2 \tag{1.1}$$

It is noteworthy that d varies with machine and the technique used. Variation in the E_D, on the other hand, directly influences the energy input to produce an AM product. Consequently, the microstructure and mechanical properties of the AM products are tailored. A combination of P and s which yields lower E_D can develop porosity in the materials and vice versa [22,23]. Other parameters that marginally or significantly influence the mechanical properties are discussed in the subsequent sections. In many cases, for instance, AM for NiTi alloys, linear energy density (E_L) is used to account for the combined effect of P and s, as shown in equation (1.2) [24].

$$E_L = \frac{P}{s} \quad J/mm \tag{1.2}$$

E_L determines the size of the melt pool during the process. An optimized linear energy density helps in identifying the shape and size of the melt pool. At a lower linear energy density, with increasing P, the melt pool size increases at constant s and vice versa [24]. The size of the melt pool signifies the area of melt zone which correspondingly alters the microstructure and mechanical properties of the materials. In fact, a combination of low P and high s also deteriorates the surface quality of the products, as compared to that for high P and low s.

6 Additive Manufacturing

1.3.1 Laser power (P) and scan speed (s)

Laser power is the source of energy that is primarily used to melt the metal powder in the AM process. The source of laser power, however, may vary depending on the types of powder to be melted and is consequently related to the method of deposition [25]. These lasers are classified based on the working principles, power output, efficiency, etc. Most common types of laser are CO_2 laser, Nd:YAG laser, Yb-fiber laser, etc. Scan speed is the speed at which the laser moves to melt the metal powder. The combined effect of s and P accounts for the microstructural change to a great extent, as discussed in the preceding section. P provides the energy to melt the metal powder. High P, therefore, ensures proper melting and bonding of the metal powders. Faster s leads to finer microstructure due to rapid cooling; consequently, mechanical properties of the materials improve correspondingly. However, faster s also leads to inappropriate or incomplete melting. That, in turn, assists in generating porosity in the as-fabricated component. Development of porosity, i.e., defects in the structure, significantly influences the mechanical properties as well [23]. Elastic modulus of the materials is also found to decrease with increasing porosity. Hence, a complex underplay of the different factors is noted to affect the microstructural evolution and mechanical properties of additively manufactured materials. Considering this, an optimized combination of properties is preferred to attain the targeted set of properties.

A combined effect of P and s can be determined by E_D, as discussed in Section 1.1 by equation (1.1). By controlling the beam size or laser diameter (d), P and s, different size and geometry of melt pools can be obtained. Such changes in these process parameters significantly influence the microstructural evolution and therefore the mechanical properties of the manufactured materials. A lower P also causes insufficient melting of the powder, which results in induced porosity. Consequently, the density of the manufactured parts also decreases. Conversely, sufficiently high P is prone to eliminate the manufacturing defects such as pores and unmelted loose powder. Accordingly, a highly dense and defect-free product can be obtained. It is noteworthy that microstructural and phase modification, if any, owing to variation in laser power does showcase the change in the mechanical properties, in turn. In addition to P, different materials show different microstructures, and hence, their response toward mechanical properties changes accordingly.

Studies by Greco et al. revealed that with increasing P, microhardness of stainless steel increases [15]. Zheng et al. reported an increase in the strength and ductility, whereas a nominal variation in the hardness is observed with increasing P [26]. However, it is noteworthy that optimization of P and s in terms of E_D is more important to estimate the overall mechanical properties. Several researches reveal the influence of the combined effect of P and s. Equation (1.1) dictates the inverse relation of P and

s with E_D. However, with an increase in P and reduction in s, substantial enhancement in the melt pool size is attained [27]. Increasing P at constant s increases E_D. Conversely, at a constant P, E_D decreases with increasing s. It is specified by Liu et al. that for the value of E_D varying within the range of ~83 to ~110 J/mm, the tensile strength of steel has an inverse relation with E_D. However, a direct relation holds for ductility [23].

Hardness and elastic modulus of *NiTi* increase with increasing P, whereas tensile strength decreases [8,9,28]. However, a reverse trend is observed for strength, which decreases with increasing P [8]. The transformation temperature of NiTi alloy increases significantly with increasing E_D. Such variation in transformation temperature is attributed to the microstructural refinement and phase change [8]. Rapid cooling involved in the AM process is prone to develop internal stresses along with generation of dislocation and point defects. All these, on the other hand, act in restricting the reverse transformation of NiTi. Consequently, the reverse martensitic transformation temperature increases. Speirs et al. reported a decrease in martensitic transformation temperature with increasing s. This, however, is related to the compositional change owing to evaporation of nickel during the AM process [21].

1.3.2 Hatch spacing (h) and layer thickness (t)

Hatch spacing, h, signifies the distance between two successive neighboring laser movements. On the other hand, layer thickness, t, is basically the distance between two successive layers, as illustrated in Figures 1.1 and 1.2. h and t are also the most crucial parameters that dictate microstructure generation and hence mechanical properties of the AM product. Effect of h on the deposited layer is illustrated in Figure 1.2. Many a time, h and t are also included to estimate E_D, as shown in equation (1.3) [24,29]. In fact, t also maintains a relation with P and s.

$$E_D = \frac{P}{s \times h \times t} \quad J/mm^2 \tag{1.3}$$

Altering any or all of the parameters, namely, P, h, s and t, can lead to a variation in the values of E_D.

It can be seen that with decreasing h, overlapping of the layer increases. This overlapping, on the other hand, leads to reheating/remelting of the successive layers. In fact, it is noteworthy that each layer gets reheated, while the successive top layer is deposited. Shorter h suffers from frequent reheating/remelting, which alters the melt pool size and in turn varies the microstructural evolution as well. Significant microstructural changes are noticed in such cases in comparison to the ones devoid of overlapping [30]. A systematic study by Ma et al. revealed that the overlapped layered microstructure has lower dislocation density as compared to the layer without

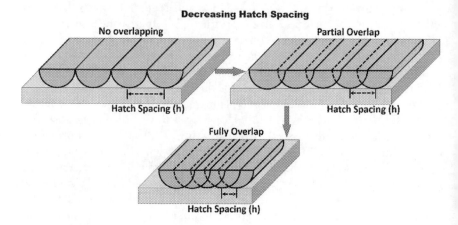

Figure 1.2 Hatch spacing illustrating the overlapping of successive layers.

overlapping [31]. This might be because the reheating of the successive layer causes annihilation or recovery of the dislocation. It is noteworthy that the presence of dislocation significantly changes the mechanical properties of the materials, and hence, it is expected that with decreasing h, lower hardness and residual stress can be observed.

Stainless steel components that are additively manufactured using high value of E_D yield higher hardness as compared to those that have been fabricated with lower E_D. Hardness values are also noted to vary directly with t [15]. On the contrary, Rońda et al. reported a reduction in the tensile strength for AM deposited steel with increasing t [32]. Nevertheless, some studies do report only a nominal change in the hardness and tensile strength with increasing t [33,34]. Such marginal differences in the mechanical properties of the materials with t are attributed to the subtle variation in microstructure between the two. An enhancement in the depth of the melt pool with corresponding reduction in its width is also noted with an increase in h [35].

In the case of NiTi-based materials, it is noted by Saghaian et al. [24] that h maintains a direct relationship with the porosity level of the additively manufactured alloy and consequently affects the mechanical properties. In fact, a shorter h leads to overlap and remelting of the previously deposited layer and hence shows a different microstructure for the remelted zone. It is observed that within a melt pool, different morphologies of grains and microstructure can be developed, which further affect the mechanical behavior of the materials. A systematic microscale indentation–based study on individual grains by Kumar et al. on additively manufactured NiTi confirmed that finer grains yield higher hardness at the small scale as compared to that obtained from the larger grains [12].

1.3.3 Scan orientation (θ) and test orientation (θ')

Scan orientation (θ) is the direction/scan difference between two successive layers, as illustrated in Figure 1.3a. Test orientation (θ'), on the other hand, is the orientation of the loading axis with respect to the build direction, as shown in Figure 1.3b. θ may be similar with each successive layer. In other words, the scanning pattern is the same on each layer or varies with some angle with the successive layer. The scanning pattern is repeated in intervals to achieve fully built components. For example, the schematic shown in Figure 1.3a represents a scenario where the lower layer has an orientation difference of 90° with respect to the top layer. It is noteworthy that the materials manufactured with an orientation of 90° with respect to the successive layer will result in a similar morphology, parallel (x–y plane and y–z plane) to the build direction (see Figure 1.3). Conversely, the microstructure will be different in the plane perpendicular to the build direction. However, the other values of θ may lead to different morphologies on either side.

Research efforts by various scientific groups revealed that irrespective of the materials, variation in θ for AM leads to notable modification in the mechanical properties [36–38]. The θ may vary from a unidirectional scan at each successive layer to some complicated scanning patterns, i.e., scan at some angle with respect to the preceding layer. Larimian et al. studied the effect of θ on the variation in the density of additively manufactured 316L stainless steel, which indirectly affects the mechanical properties. Interestingly, unidirectional scan is noted to yield a better densification of ~99.4%. On the other hand, the complex scanning patterns yield approximately 97%–98% densification due to higher cooling rates. This, however, results in refined microstructures [37]. A similar observation was established by Leicht et al. on densification of this category of stainless steel. However, a reduction in the hardness is also reported [36]. Song et al. noted improved performance with 47° layer orientation for additively

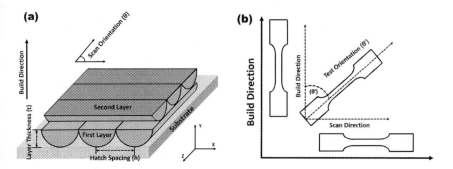

Figure 1.3 (a) Schematic illustrating the layer orientation of two consecutive layers. (b) Illustration of test orientation.

10   Additive Manufacturing

manufactured stainless steel [39]. However, stainless steel components, manufactured with θ' 90°, revealed better mechanical properties as compared to that for θ' being parallel to the build direction [36].

It is revealed that tensile properties are better in the direction parallel to the build direction as compared to the perpendicular direction [40]. On the contrary, some researchers reported a reverse trend in tensile properties. They revealed that specimens fabricated along the perpendicular direction yield improved tensile strength, irrespective of the deposited orientation [41–47]. On the other hand, Jaydeep et al. reported that 45° orientation yields better tensile strength. However, they have also pointed out that the variation in the tensile behavior is only nominal [48]. In a different observation, where specimens are deposited in the edge direction $(x-y)$, i.e., the specimens are perpendicular to the build direction and their thickness is in the corresponding z-direction, better tensile strength and ductility are realized [49,50]. Overall, the studies show diversified estimation of the mechanical properties. However, the variation on the mechanical properties is only nominal and particularly depends on the microstructural evolution during deposition and processing methods. It has been found that finer equiaxed grain results in better tensile properties with respect to fine columnar microstructure. Some research efforts also revealed that there are only nominal or negligible effects of orientation on the mechanical properties of additively manufactured materials [51].

Moghaddam et al. studied the effect of θ' on the tensile behavior of NiTi-based shape memory alloy. They reported better tensile properties with the loading axis being perpendicular to the build direction and vice versa [52]. θ of 45° is noted to significantly reduce the shape recovery characteristics. No evidence of change in the transformation temperature due to change in the scanning orientation is available. However, stress-induced martensitic transformation (SIMT) temperature increases with increasing tensile strength [53]. Nematollahi et al. observed a similar linear relation with the SIMT [38]. Nevertheless, it is apparent from the detailed reviews that the overall effect of the θ' on the transformation temperature requires further systematic investigation.

1.3.4 Deposition depth (H)

Overall, the deposition depth (H) corresponds to the total thickness of the deposited product. It is observed that the microstructure of the material is found to be different at each successive layer of deposition due to corresponding heat treatments at each layer [54]. Consequently, mechanical properties of the materials also vary. The initial layer adjacent to/in contact with the surface of the substrate undergoes rapid cooling. Hence, in the vicinity of the substrate surface, fine equiaxed grains are observed [55].

Further deposition of the layer leads to repeated heating of the successive layers. Correspondingly, grain growth is observed owing to aging and comparatively coarser grains form with increasing distance from the substrate. On the contrary, the top most layer of the product, being the last layer, is devoid of any such reheating. Furthermore, the top most layer is in contact with the environment, thereby leading to a higher cooling rate. Hence, the surface layer is prone to generate finer grains. Since mechanical properties of materials are primarily controlled by the microstructural size, it is noted to vary with the thickness of the product. The primary tool to characterize the mechanical behavior of materials is indentation. Depth dependence mechanical properties of the materials are characterized by many researchers. The initial indentation behavior of the materials shows that hardness varies from top to bottom of the deposition. The lower layer (near to the substrate) shows higher hardness as compared to the middle and top portions of the deposition [54,56].

Wang et al. studied the depth (H)-dependent mechanical properties of NiTi alloy manufactured through the arc welding AM process. They reported that hardness values increase with an increase in H. In other words, the bottom layer exhibits the lowest hardness as compared to the top layer [57]. A similar trend was observed with tensile strength. However, a reverse trend on the transformation temperature was observed where the lowest transformation temperature was noted at the top layer. Such variation in the mechanical properties and transformation behavior of NiTi was attributed to the presence of Ni_4Ti_3 precipitates. Therefore, the depth dependence variation of mechanical properties is dependent on the microstructure and phase evolution during the heating/reheating schedule during AM.

Overall, the role of different AM process parameters in influencing the mechanical properties of metallic systems, particularly a commonly used material, stainless steel and a unique category of material, NiTi-based shape memory alloy, is thoroughly discussed in this chapter.

1.4 CASE STUDY

Observations related to the effect of process parameters on the microstructure and mechanical properties of additively manufactured alloys are highlighted in this section. Representative alloys from both the non-ferrous and ferrous categories are chosen to develop a broader insight. While NiTi-based shape memory alloy serves the first category, 316L stainless steel is considered for the second one.

NiTi shape memory alloys: Interesting observations made upon investigating the role of varying AM process parameters on the microstructure evolution and mechanical properties of laser engineered net shaping

(LENS)-based NiTi alloy are discussed here [12]. Different processing parameters (laser power and scanning speed) corresponding to varying laser energy densities, viz. E_D of 27, 40, and 70 J/mm^2, are considered. It is reported that with variation of E_D, the microstructure, porosity as well as hardness, strength and strain recoverability of NiTi alloys alter significantly. The microstructure of the additively manufactured alloy does reveal typical signatures with pronounced hatch spacing and layer thickness. Nevertheless, the one manufactured with the lowest E_D particularly revealed three distinct zones in the microstructure including fine equiaxed, columnar, and coarse equiaxed grains, as shown in Figure 1.4a. With increasing E_D, the porosity level in the alloy is, however, noted to reduce. This is particularly related to the fact that sufficiently high laser power facilitates appropriate melting of the metal powder. Comparatively higher hardness is also observed for the NiTi alloy manufactured with the highest E_D. Such variation in the microstructure and mechanical properties with increasing E_D corresponds to generating precipitates with increasing laser power. Nevertheless, the most relevant mechanical property of interest for NiTi is pseudoelasticity or shape recovery, rather than strength. Interestingly, it is realized that perfect pseudoelasticity for this functional material can only be attained while the NiTi alloy is manufactured with $E_D = 40$ J/mm^2. A similar trend is also evident when tested at a small scale using a spherical nanoindenter. The role of process parameters in controlling the required performance of a specialized functional material is thereby apparent from this study.

316L stainless steel: The evolution in microstructure and mechanical properties of laser metal deposition (LMD) manufactured stainless steel of grade 316L has been systematically studied [54]. Microstructure of the alloy, manufactured with $E_D = 33.33$ J/mm^2, does reveal the typical signatures of AM. Nevertheless, both the size and morphology of the salient microstructural features are noted to vary with distance from the substrate, along the build direction. This is realized upon characterizing the different sections of the as-build specimen, as illustrated in Figure 1.4b. Along with the presence of melt pools, the distinct microstructural features with columnar and equiaxed grains are also noted in the as-built 316L alloy, as shown Figure 1.4c and d. A finer microstructure was observed at the bottom section, while it is observed to coarsen toward the top layer. Fast cooling at the bottom section leads to grain refinement. A corresponding effect in modifying the properties is also investigated in detail. The microhardness of 316L was found to be high at the bottom, owing to finer microstructure, and it decreases toward the top layers.

Both of these studies on non-ferrous and ferrous alloys do signify a common fact that the process of AM as well as any variation in the corresponding process parameters significantly modifies the microstructure and mechanical properties. Optimized parameters should be employed during the AM process to achieve the best combination of microstructural and mechanical properties that are relevant for specific applications.

Effect of process parameters on mechanical properties 13

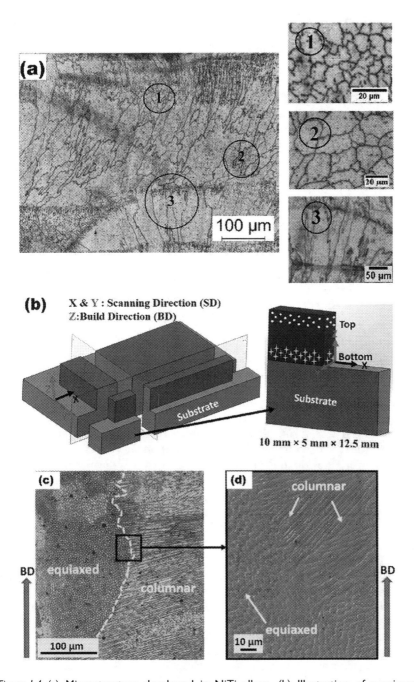

Figure 1.4 (a) Microstructure developed in NiTi alloys. (b) Illustration of specimen geometry for microstructural and mechanical properties characterization of the stainless steel 316L. (c) Microstructure developed in 316L at low magnification. (d) Microstructure of 316L at high magnification to reveal equiaxed and columnar grains.

1.5 LIMITATION AND FUTURE SCOPE

A detailed and extensive literature review clarifies that different process parameters alter the mechanical properties of the NiTi and stainless steel. A paradox in the observations is also noted among the various attempts by the different scientific groups, particularly related to the different methodologies adopted and the different objectives planned. Such ambiguities are highlighted in the chapter. Furthermore, it is also realized that the performance of any specific alloy system manufactured through any particular additive process still depends on the process parameters. Obtaining a particular set of microstructures and thereby properties therefore warrants implication of optimized process parameters for AM. This mandates further systematic research to be carried out in this regard.

1.6 CLOSURE

A comprehensive literature survey did highlight that the process parameters significantly affect the overall mechanical properties of the materials fabricated using AM. The primary modification is reflected in the microstructural features and by optimization of the AM process parameters; these features can be tailored further. Faster scan speed in combination with lower laser power leads to porosity while manufacturing the component. Simultaneously, hatch spacing and layer thickness also modify the density of the materials and thereby improve the mechanical properties of the fabricated materials. A significant alteration in the mechanical properties is also observed by varying the scan strategies and test orientations. While unidirectional scan strategies densify the component, the one involving perpendicular orientation to the build direction enhances the mechanical properties. Most importantly, optimized combination of process parameters enables additively manufacturing a metallic system with desired mechanical properties by engineering the microstructure of the product.

REFERENCES

1. I. Gibson, D.W. Rosen, B. Stucker, *Additive Manufacturing Technologies: Rapid Prototyping to Direct Digital Manufacturing*, 2009. https://doi.org/10.1007/978-1-4419-1120-9.
2. F. Calignano, D. Manfredi, E.P. Ambrosio, S. Biamino, M. Lombarda, E. Atzeni, A. Salmi, P. Minetola, L. Iuliano, P. Fino, Overview on additive manufacturing technologies, *Proc. IEEE.* 105 (2017) 593–612. https://doi.org/10.1109/JPROC.2016.2625098.
3. D. Herzog, V. Seyda, E. Wycisk, C. Emmelmann, Additive manufacturing of metals, *Acta Mater.* 117 (2016) 371–392. https://doi.org/10.1016/j.actamat.2016.07.019.

4. W.E. Frazier, Metal additive manufacturing: A review, *J. Mater. Eng. Perform.* 23 (2014) 1917–1928. https://doi.org/10.1007/s11665-014-0958-z.
5. ASTM-F2792-12a, Standard terminology for additive manufacturing technologies, *ASTM Int.* (2013) 10–12. https://doi.org/10.1520/F2792-12A.2.
6. C.Y. Yap, C.K. Chua, Z.L. Dong, Z.H. Liu, D.Q. Zhang, L.E. Loh, S.L. Sing, Review of selective laser melting: Materials and applications, *Appl. Phys. Rev.* 2(041101) (2015) 1–21. https://doi.org/10.1063/1.4935926.
7. H. Galarraga, R.J. Warren, D.A. Lados, R.R. Dehoff, M.M. Kirka, P. Nandwana, Effects of heat treatments on microstructure and properties of Ti-6Al-4V ELI alloy fabricated by electron beam melting (EBM), *Mater. Sci. Eng. A.* 685 (2017) 417–428. https://doi.org/10.1016/j.msea.2017.01.019.
8. B.V. Krishna, S. Bose, A. Bandyopadhyay, Laser processing of net-shape NiTi shape memory alloy, *Metall. Mater. Trans. A.* 38 (2007) 1096–1103. https://doi.org/10.1007/s11661-007-9127-4.
9. B.V. Krishna, S. Bose, A. Bandyopadhyay, Fabrication of porous NiTi shape memory alloy structures using laser engineered net shaping, *J. Biomed. Mater. Res. – Part B Appl. Biomater.* 89 (2009) 481–490. https://doi.org/10.1002/jbm.b.31238.
10. A. Bandyopadhyay, B.V. Krishna, W. Zue, S. Bose, Application of laser engineered net shaping (LENS) to manufacture porous and functionally graded structures for load bearing implants, *J. Mater. Sci. Mater. Med.* 20 (2009) 29–34. https://doi.org/10.1007/s10856-008-3478-2.
11. A. Baran, M. Polanski, Microstructure and properties of LENS (laser engineered net shaping) manufactured Ni-Ti shape memory alloy, *J. Alloys Compd.* 750 (2018) 863–870. https://doi.org/10.1016/j.jallcom.2018.03.400.
12. S. Kumar, L. Marandi, V.K. Balla, S. Bysakh, D. Piorunek, G. Eggeler, M. Das, I. Sen, Microstructure – Property correlations for additively manufactured NiTi based shape memory alloys, *Materialia.* 8 (2019) 100456. https://doi.org/10.1016/j.mtla.2019.100456.
13. M.A. Melia, J.D. Carroll, S.R. Whetten, S.N. Esmaeely, J. Locke, E. White, I. Anderson, M. Chandross, J.R. Michael, N. Argibay, E.J. Schindelholz, A.B. Kustas, Mechanical and corrosion properties of additively manufactured CoCrFeMnNi high entropy alloy, *Addit. Manuf.* 29 (2019). https://doi.org/10.1016/j.addma.2019.100833.
14. W.U.H. Syed, A.J. Pinkerton, L. Li, A comparative study of wire feeding and powder feeding in direct diode laser deposition for rapid prototyping, *Appl. Surf. Sci.* 247 (2005) 268–276. https://doi.org/10.1016/j.apsusc.2005.01.138.
15. S. Greco, K. Gutzeit, H. Hotz, B. Kirsch, J.C. Aurich, Selective laser melting (SLM) of AISI 316L—Impact of laser power, layer thickness, and hatch spacing on roughness, density, and microhardness at constant input energy density, *Int. J. Adv. Manuf. Technol.* 108 (2020) 1551–1562. https://doi.org/10.1007/s00170-020-05510-8.
16. I. Shishkovsky, F. Missemer, I. Smurov, Direct metal deposition of functional graded structures in Ti-Al system, *Phys. Procedia.* 39 (2012) 382–391. https://doi.org/10.1016/j.phpro.2012.10.052.
17. K.P. Karunakaran, S. Suryakumar, V. Pushpa, S. Akula, Low cost integration of additive and subtractive processes for hybrid layered manufacturing, *Robot. Comput. Integr. Manuf.* 26 (2010) 490–499. https://doi.org/10.1016/j.rcim.2010.03.008.

18. D. Ding, Z. Pan, D. Cuiuri, H. Li, Wire-feed additive manufacturing of metal components: Technologies, developments and future interests, *Int. J. Adv. Manuf. Technol.* 81 (2015) 465–481. https://doi.org/10.1007/s00170-015-7077-3.

19. W.U.H. Syed, A.J. Pinkerton, L. Li, Combining wire and coaxial powder feeding in laser direct metal deposition for rapid prototyping, *Appl. Surf. Sci.* 252 (2006) 4803–4808. https://doi.org/10.1016/j.apsusc.2005.08.118.

20. J. Frenzel, E.P. George, A. Dlouhy, C. Somsen, M.F.-X. Wagner, G. Eggeler, Influence of Ni on martensitic phase transformations in NiTi shape memory alloys, *Acta Mater.* 58 (2010) 3444–3458. https://doi.org/10.1016/j.actamat.2010.02.019.

21. M. Speirs, X. Wang, S. Van Baelen, A. Ahadi, S. Dadbakhsh, J.P. Kruth, J. Van Humbeeck, On the transformation behavior of NiTi shape-memory alloy produced by SLM, *Shape Mem. Superelasticity.* 2 (2016) 310–316. https://doi.org/10.1007/s40830-016-0083-y.

22. S. Saedi, N. Shayesteh Moghaddam, A. Amerinatanzi, M. Elahinia, H.E. Karaca, On the effects of selective laser melting process parameters on microstructure and thermomechanical response of Ni-rich NiTi, *Acta Mater.* 144 (2018) 552–560. https://doi.org/10.1016/j.actamat.2017.10.072.

23. J. Liu, Y. Song, C. Chen, X. Wang, H. Li, C. Zhou, J. Wang, K. Guo, J. Sun, Effect of scanning speed on the microstructure and mechanical behavior of 316L stainless steel fabricated by selective laser melting, *Mater. Des.* 186 (2020). https://doi.org/10.1016/j.matdes.2019.108355.

24. S. Ehsan Saghaian, M. Nematollahi, G. Toker, A. Hinojos, N. Shayesteh Moghaddam, S. Saedi, C.Y. Lu, M. Javad Mahtabi, M.J. Mills, M. Elahinia, H.E. Karaca, Effect of hatch spacing and laser power on microstructure, texture, and thermomechanical properties of laser powder bed fusion (L-PBF) additively manufactured NiTi, *Opt. Laser Technol.* 149 (2022) 107680. https://doi.org/10.1016/j.optlastec.2021.107680.

25. H. Lee, C.H.J. Lim, M.J. Low, N. Tham, V.M. Murukeshan, Y.J. Kim, Lasers in additive manufacturing: A review, *Int. J. Precis. Eng. Manuf. – Green Technol.* 4 (2017) 307–322. https://doi.org/10.1007/s40684-017-0037-7.

26. Z. Zheng, L. Wang, B. Yan, Effects of laser power on the microstructure and mechanical properties of 316L stainless steel prepared by selective laser melting, *Int. J. Mod. Phys. B.* 31 (2017) 1–5. https://doi.org/10.1142/S0217979217440155.

27. J. Kang, J. Yi, T. Wang, X. Wang, T. Feng, Y. Feng, P. Wu, Effect of laser power and scanning speed on the microstructure and mechanical properties of SLM fabricated Inconel 718 specimens, *Mater. Sci. Eng. Int. J.* 3 (2019) 72–76. https://doi.org/10.15406/mseij.2019.03.00094.

28. J.J. Marattukalam, A. Kumar, S. Datta, M. Das, V. Krishna, S. Bontha, S.K. Kalpathy, Microstructure and corrosion behavior of laser processed NiTi alloy, *Mater. Sci. Eng. C.* 57 (2015) 309–313. https://doi.org/10.1016/j.msec.2015.07.067.

29. N. Ahmed, I. Barsoum, G. Haidemenopoulos, R.K.A. Al-Rub, Process parameter selection and optimization of laser powder bed fusion for 316L stainless steel: A review, *J. Manuf. Process.* 75 (2022) 415–434. https://doi.org/10.1016/j.jmapro.2021.12.064.

30. B. Feng, C. Wang, Q. Zhang, Y. Ren, L. Cui, Q. Yang, S. Hao, Effect of laser hatch spacing on the pore defects, phase transformation and properties of selective laser melting fabricated NiTi shape memory alloys, *Mater. Sci. Eng. A.* 840 (2022) 142965. https://doi.org/10.1016/j.msea.2022.142965.

31. J. Ma, B. Franco, G. Tapia, K. Karayagiz, L. Johnson, J. Liu, R. Arroyave, I. Karaman, A. Elwany, Spatial control of functional response in 4D-printed active metallic structures, *Sci. Rep.* 7 (2017) 1–8. https://doi.org/10.1038/srep46707.

32. N. Rońda, K. Grzelak, M. Polański, J. Dworecka-Wójcik, The influence of layer thickness on the microstructure and mechanical properties of M300 maraging steel additively manufactured by LENS® technology, *Materials (Basel).* 15 (2022). https://doi.org/10.3390/ma15020603.

33. S.A.R. Shamsdini, S. Shakerin, A. Hadadzadeh, B.S. Amirkhiz, M. Mohammadi, A trade-off between powder layer thickness and mechanical properties in additively manufactured maraging steels, *Mater. Sci. Eng. A.* 776 (2020) 139041. https://doi.org/10.1016/j.msea.2020.139041.

34. H.C. Hyer, C.M. Petrie, Effect of powder layer thickness on the microstructural developmet on additively manufactured SS316, *J. Manuf. Process.* 76 (2022) 666–674.

35. Z. Dong, Y. Liu, W. Wen, J. Ge, J. Liang, Effect of hatch spacing on melt pool and as-built quality during selective laser melting of stainless steel: Modeling and experimental approaches, *Materials (Basel).* 12 (2018) 1–15. https://doi.org/10.3390/ma12010050.

36. A. Leicht, C.H. Yu, V. Luzin, U. Klement, E. Hryha, Effect of scan rotation on the microstructure development and mechanical properties of 316L parts produced by laser powder bed fusion, *Mater. Charact.* 163 (2020) 2–10. https://doi.org/10.1016/j.matchar.2020.110309.

37. T. Larimian, M. Kannan, D. Grzesiak, B. AlMangour, T. Borkar, Effect of energy density and scanning strategy on densification, microstructure and mechanical properties of 316L stainless steel processed via selective laser melting, *Mater. Sci. Eng. A.* 770 (2020) 138455. https://doi.org/10.1016/j.msea.2019.138455.

38. M. Nematollahi, S.E. Saghaian, K. Safaei, P. Bayati, P. Bassani, C. Biffi, A. Tuissi, H. Karaca, M. Elahinia, Building orientation-structure-property in laser powder bed fusion of NiTi shape memory alloy, *J. Alloys Compd.* 873 (2021) 159791. https://doi.org/10.1016/j.jallcom.2021.159791.

39. Y. Song, Q. Sun, K. Guo, X. Wang, J. Liu, J. Sun, Effect of scanning strategies on the microstructure and mechanical behavior of 316L stainless steel fabricated by selective laser melting, *Mater. Sci. Eng. A.* 793 (2020) 139879. https://doi.org/10.1016/j.msea.2020.139879.

40. I. Hacısalihoğlu, F. Yıldız, A. Çelik, The effects of build orientation and hatch spacing on mechanical properties of medical Ti–6Al–4V alloy manufactured by selective laser melting, Mater. *Sci. Eng. A.* 802 (2021) 140649. https://doi.org/10.1016/j.msea.2020.140649.

41. T. Pan, X. Zhang, A. Flood, S. Karnati, W. Li, J. Newkirk, F. Liou, Effect of processing parameters and build orientation on microstructure and performance of AISI stainless steel 304L made with selective laser melting under different strain rates, *Mater. Sci. Eng. A.* 835 (2022) 142686. https://doi.org/10.1016/j.msea.2022.142686.

42. G. Sander, A.P. Babu, X. Gao, D. Jiang, N. Birbilis, On the effect of build orientation and residual stress on the corrosion of 316L stainless steel prepared by selective laser melting, *Corros. Sci.* 179 (2021) 109149. https://doi.org/10.1016/j.corsci.2020.109149.

43. Z. Xie, Y. Dai, X. Ou, S. Ni, M. Song, Effects of selective laser melting build orientations on the microstructure and tensile performance of Ti–6Al–4V alloy, *Mater. Sci. Eng. A.* 776 (2020) 139001. https://doi.org/10.1016/j.msea.2020.139001.

44. P. Hartunian, M. Eshraghi, Effect of build orientation on the microstructure and mechanical properties of selective laser-melted Ti-6Al-4V Alloy, *J. Manuf. Mater. Process.* 2 (2018). https://doi.org/10.3390/jmmp2040069.

45. M. Mukherjee, Effect of build geometry and orientation on microstructure and properties of additively manufactured 316L stainless steel by laser metal deposition, *Materialia.* 7 (2019) 100359. https://doi.org/10.1016/j.mtla.2019.100359.

46. Y.T. Tang, J.E. Campbell, M. Burley, J. Dean, R.C. Reed, T.W. Clyne, Profilometry-based indentation plastometry to obtain stress-strain curves from anisotropic superalloy components made by additive manufacturing, *Materialia.* 15 (2021) 101017. https://doi.org/10.1016/j.mtla.2021.101017.

47. S. Pal, N. Gubeljak, R. Hudak, G. Lojen, V. Rajtukova, J. Predan, V. Kokol, I. Drstvensek, Tensile properties of selective laser melting products affected by building orientation and energy density, *Mater. Sci. Eng. A.* 743 (2019) 637–647. https://doi.org/10.1016/j.msea.2018.11.130.

48. J. Vishwakarma, K. Chattopadhyay, N.C. Santhi Srinivas, Effect of build orientation on microstructure and tensile behaviour of selectively laser melted M300 maraging steel, *Mater. Sci. Eng. A.* 798 (2020) 140130. https://doi.org/10.1016/j.msea.2020.140130.

49. M. Simonelli, Y.Y. Tse, C. Tuck, Effect of the build orientation on the mechanical properties and fracture modes of SLM Ti-6Al-4V, *Mater. Sci. Eng. A.* 616 (2014) 1–11. https://doi.org/10.1016/j.msea.2014.07.086.

50. E. Pehlivan, M. Roudnicka, J. Dzugan, M. Koukolikova, V. Králík, M. Seifi, J.J. Lewandowski, D. Dalibor, M. Daniel, Effects of build orientation and sample geometry on the mechanical response of miniature CP-Ti Grade 2 strut samples manufactured by laser powder bed fusion, *Addit. Manuf.* 35 (2020). https://doi.org/10.1016/j.addma.2020.101403.

51. G. Meneghetti, D. Rigon, D. Cozzi, W. Waldhauser, M. Dabalà, Influence of build orientation on static and axial fatigue properties of maraging steel specimens produced by additive manufacturing, *Procedia Struct. Integr.* 7 (2017) 149–157. https://doi.org/10.1016/j.prostr.2017.11.072.

52. N. Shayesteh Moghaddam, S. Saedi, A. Amerinatanzi, A. Hinojos, A. Ramazani, J. Kundin, M.J. Mills, H. Karaca, M. Elahinia, Achieving superelasticity in additively manufactured NiTi in compression without postprocess heat treatment, *Sci. Rep.* 9 (2019) 1–11. https://doi.org/10.1038/s41598-018-36641-4.

53. N. Shayesteh Moghaddam, S.E. Saghaian, A. Amerinatanzi, H. Ibrahim, P. Li, G.P. Toker, H.E. Karaca, M. Elahinia, Anisotropic tensile and actuation properties of NiTi fabricated with selective laser melting, *Mater. Sci. Eng. A.* 724 (2018) 220–230. https://doi.org/10.1016/j.msea.2018.03.072.

54. P. Nath, D. Nanda, G.P. Dinda, I. Sen, Assessment of microstructural evolution and mechanical properties of laser metal deposited 316L stainless steel, *J. Mater. Eng. Perform.* 30 (2021) 6996–7006. https://doi.org/10.1007/s11665-021-06101-8.
55. G.F. Sun, X.T. Shen, Z.D. Wang, M.J. Zhan, S. Yao, R. Zhou, Z.H. Ni, Laser metal deposition as repair technology for 316L stainless steel: Influence of feeding powder compositions on microstructure and mechanical properties, *Opt. Laser Technol.* 109 (2019) 71–83. https://doi.org/10.1016/j.optlastec.2018.07.051.
56. I. Avula, A.C. Arohi, C.S. Kumar, I. Sen, Microstructure, corrosion and mechanical behavior of 15-5 PH stainless steel processed by direct metal laser sintering, *J. Mater. Eng. Perform.* 30 (2021) 6924–6937. https://doi.org/10.1007/s11665-021-06069-5.
57. J. Wang, Z. Pan, G. Yang, J. Han, X. Chen, H. Li, Location dependence of microstructure, phase transformation temperature and mechanical properties on Ni-rich NiTi alloy fabricated by wire arc additive manufacturing, *Mater. Sci. Eng. A.* 749 (2019) 218–222. https://doi.org/10.1016/j.msea.2019.02.029.

Chapter 2

Parametric study of fused deposition modelling

Kriti Srivastava and Yogesh Kumar
National Institute of Technology (Patna)

CONTENTS

2.1	Introduction	21
	2.1.1 Additive manufacturing	21
	2.1.2 Fused deposition modelling	23
2.2	Energy consumption	24
2.3	Process parameters	25
	2.3.1 Raster width	25
	2.3.2 Raster angle	26
	2.3.3 Layer height	27
	2.3.4 Infill pattern and percentage	28
	2.3.5 Build orientation	30
	2.3.6 Air gap	31
	2.3.7 Extrusion temperature	32
	2.3.8 Build time	32
	2.3.9 Materials in fused deposition modelling	35
	2.3.10 Application	35
2.4	Post-processing techniques	35
2.5	Limitations and challenges	36
2.6	Conclusion and future scope	37
References		37

2.1 INTRODUCTION

2.1.1 Additive manufacturing

Additive manufacturing (AM) is a well-thought-out manufacturing method that includes layering materials to produce three-dimensional (3D) parts directly from computer aided design (CAD) models (Hegab 2016). The primary advantages of this approach are its capacity to produce nearly any profile that would not be possible to produce by any other manufacturing process. Other advantages are reduced time and cost, amplified human engagement, and, as an outcome, a more rapid product development cycle

DOI: 10.1201/9781003258391-2

22 Additive Manufacturing

(Chaunier et al. 2018). Each design and development procedure utilising a 3D printing machine necessitates the operator performing a specific set of tasks. The accessibility of such a task sequence is emphasised by convenient 3D printing devices. These desk machines are distinguished by the low price, ease of usage, and capability to be used in an industrial and academic environment. Each stage of this system is expected to have a restricted number of alternatives and requires the least effort. Nevertheless, this implies that there are restricted options, like restricted selection of materials and many other parameters with which to do experimentation. Larger, more versatile machines are more capable of being adjusted to meet a wide range of user needs, but they are also more difficult to manage. Types of Additive manufacturing and their working principles are shown in Table 2.1.

The development of AM technology over decades is shown in Figure 2.1.

Table 2.1 Types of additive manufacturing (Standard A. S. T. M. 2013)

S. no.	AM types	Technologies	Principle
1	Metal extrusion	Fused deposition modelling Contour crafting	Melted material is deposited through a nozzle. Its typical resolution is 100 μm to 1 cm.
2	Material jetting	Inkjet printing	Using piezo printing heads, droplets of photopolymers in liquid form are deposited and UV lamps are used to cure them. Its typical resolution is 10–25 μm.
3	Direct energy deposition	Laser Engineered NetShaping (LENS) Electronic BeamWelding (EBW)	Material is melted with the help of laser and deposition in a pool of melted material. Its typical resolution is 100 μm to 1 cm.
4	Sheet lamination	Laminated object manufacturing (LOM)	In this method, layers of material are fused together, with the required form carved into each form. Its typical resolution is 200–300 μm.
5	Vat polymerisation	Stereolithography	Focused energy is deposited on the surface of a liquid photopolymer. Its typical resolution is 0.1–100 μm.
6	Binder jetting	Indirect inkjet printing	Combining components in the form of powder with liquid or solid binder. Its typical resolution is 100 μm.
7	Powder bed fusion	Direct metal laser sintering Selective laser melting (SLM) Electron beam melting	Layer wise fusion of powder particles using infrared energy. Its typical resolution is 50–100 μm.

Parametric study of fused deposition modelling 23

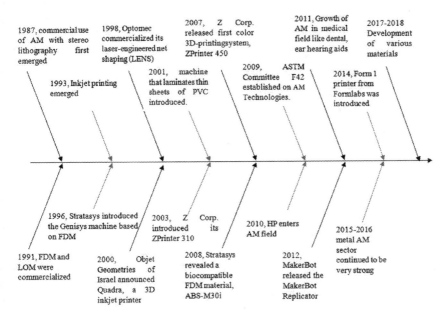

Figure 2.1 History of additive manufacturing (Wohlers & Gornet 2014).

2.1.2 Fused deposition modelling

Crump patented fused deposition modelling (FDM) was founded in 1988, and Stratasys Corporation was founded in 1989. FDM has a simple basic aspect yet can build complex geometries. As illustrated in Figure 2.2, it is an extrusion process that involves extruding melt filament feedstock through a nozzle that is controlled by an electric motor.

The filament is melted using a heated liquefier. Stepper motors move the extruder head unit across a build platform. The filament in the melted form is extruded through the heated liquefier to the nozzle, which deposits the melted filament on the build platform in the X–Y plane (worktable that is without fixture). Following the accomplishment of deposition at each subsequent cross section, the platform or print head travels down or up, respectively, by exactly one layer height along the z-axis (Galantucci 2015). As a result, a layer-by-layer method is used to construct the 3D design. This technique is repeated until the component is completed. The basic steps involved in FDM are shown in Figure 2.3.

There is a collection of material along the part's edge at the start of the Fused Deposition Modelling method, and later in the inner region of the shape.

To pack the component according to the required response, a specific set of outlines is needed. Fused Deposition Modelling printers are currently

24 Additive Manufacturing

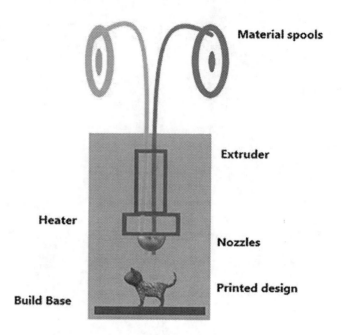

Figure 2.2 Fused deposition modelling process.

Figure 2.3 Steps in fused deposition modelling.

being used to make curved items out of thermoplastics like Polylactic acid (PLA) at a lower cost. Although the mechanical characteristics and strength of these parts are restricted, they could be improved by reinforcing the polymer with a fibre.

2.2 ENERGY CONSUMPTION

The given equation (2.1) can be utilised to model the consumption of energy of the FDM process (Haghighi & Li 2018):

$$U_{build} = U_{process} + U_{standby} + U_{heating} \qquad (2.1)$$

$U_{process}$ is the component of consumption of energy that depends on geometry, which may be determined by adding sufficient energy for up–down and left–right nozzle motions, as well as the power required for extruding the melted material, for printing the required geometry.

$U_{heating}$ is the quantity of energy necessary to warm the filament to the temperature (glass transition) and maintain it at that temperature for the entire duration of the print.

$U_{standby}$ is the element of energy that depends on time. It is the lowest quantity of energy that is essential to preserve the system on through the procedure of building, for example, energy consumption by wire, fan, etc.

2.3 PROCESS PARAMETERS

2.3.1 Raster width

The amount of melted filament extruded through the nozzle is equivalent to the amount of deposition at a similar time, as shown in Figure 2.4.

Consequently, the width of deposition ω could be evaluated by equation (2.2):

$$\pi \left(\frac{D_N}{2}\right)^2 V_E = \pi \left(\frac{\omega}{2}\right)\left(\frac{H_L}{2}\right) V_F \qquad (2.2)$$

where D_N denotes the inner diameter of the nozzle. Depending on the material loading pressure and velocity, V_E is the speed at which the string is squeezed out via the warmed injector. V_F refers to the velocity of the filling, H_L is the thickness of the layer, and ω refers to the width of the deposition (Yang et al. 2019).

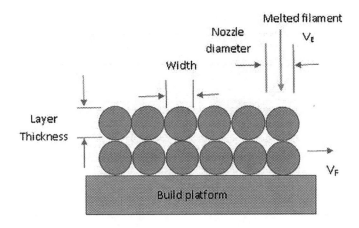

Figure 2.4 Schematic diagram of raster width.

The most noteworthy parameter influencing the quality of the surface and porosity of the component produced is observed to be the road width. The porosity decreases as the width and temperature increase.

2.3.2 Raster angle

The raster angle is well explained as the direction of the raster in relation to the stress loading direction, as shown in Figure 2.5. The material filaments are aligned parallel to the direction of the load in components produced at raster angles of 0°/90°, resulting in the strongest specimens (Wu et al. 2015). The width of the raster and its angle had the greatest impact on the flexural characteristics of ULTEM 9085 parts (Gebisa & Lemu 2018). All of the characteristics evaluated are important and have an impact, especially when the pieces are created with angle of raster other than 0. The impacts of the parameters are less evident and meaningful when the raster angle is 0°.

The angle of the raster has the greatest impact on the specimen's tensile strength. At a 0° raster angle, the strength in tensile is higher, whereas at a 90° raster angle, the strength in the tension condition is lower because the bonding zone between layer surfaces is larger at lower layer heights, and the tensile strength is higher. A larger strength in tension could be attained due to advanced raster thermal mass up to a particular level; nevertheless,

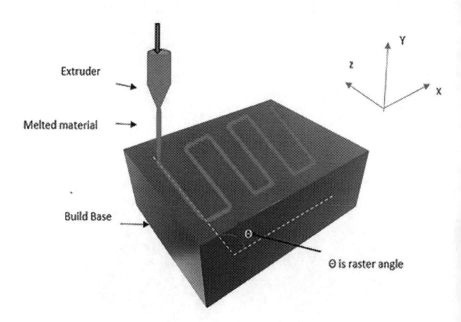

Figure 2.5 Raster angle.

formation of void between rasters occurs, and voids are a primary source of fracture beginning and propagation, weakening the parts and resulting in lower tensile strength (Rajpurohit & Dave 2018).

The interplay of the two build factors, namely, raster angle and orientation of build on solid build flexure coupons was demonstrated using a full-factorial DOE. For raised temperature flexure testing, the XYZ 0°/90° construction combination was utilised. Both yield strength and modulus of ULTEM 1010 dropped as the temperature of testing increased, as predicted, for high flexure testing up to 400°F (205°C). For sparse-build flexure forms, the XYZ 0°/90° combination of build orientation was used similarly for testing at higher temperatures (Taylor et al. 2018).

The level of infill and numbers of shells are the individual critical factors for optimising tensile characteristics, and they have to be maximised. The maximum height of layer, as well as the lowest levels of infill and quantity of shells, should be employed to optimise efficiency outputs (Griffiths et al. 2016).

2.3.3 Layer height

The thickness of the layer deposited by the extruder is a critical process factor that affects the part surface superiority. The stair-step height is determined by the thickness of the raster extruded from the nozzle or the movement in the z-axis among subsequent layers deposited on the bed, as shown in Figure 2.6.

The surface roughness increases with the increases in the road width and air gap, and decreases with the increases in the layer thickness

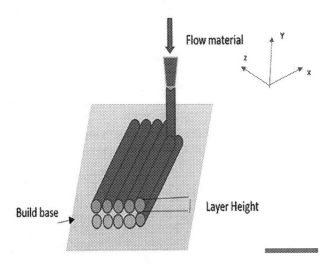

Figure 2.6 Layer height.

(Arumaikkannu et al. 2005). The reduction in the diameter of the nozzle reduces stair-stepping. It also increases production time and cost. As a result, before beginning the process, there should be a trade-off among build durations and the height of layer, as lowering the time of build might rise the stepping of stair that necessitate further post-processing (Vasudevarao et al. 2000). To use the shortest layer height possible in order to achieve good dimension accuracy and fine surface quality (Nancharaiah et al. 2010).

The range of this parameter is restricted by a minimum layer thickness beyond which it could not be reduced. Furthermore, this method is beneficial when the thickness of the layer of the FDM 3D printer may be adjusted, but very few devices offer this option, necessitating optimisation studies of a few other useful factors (Novak-Marcincin et al. 2012).

The roughness of the surface increases with increase in the thickness of the layer (Vijay et al. 2011). A reduced slice height is advised to reduce the consumption of support material, while shorter slice heights and positive air gaps are desired for minimising model material consumption (Ali et al. 2014). The viscoelastic properties are greatly influenced by the width of raster and the height of layer. The viscoelastic properties of the FDM parts were shown to be 55% influenced by the layer height and 31% influenced by the raster width (Dakshinamurthy & Gupta 2018).

2.3.4 Infill pattern and percentage

It denotes the method of printing the internal structure of the part being printed. A variety of filling patterns are accessible, including hexagonal, linear, and diamond (Qattawi et al. 2017). The frequently used infill form in Fused Deposition Modelling is the hexagonal pattern (Dave et al. 2021). Figure 2.7 shows various types of patterns used by different researchers.

The effects of infill patterns at the mechanical behaviour of printed items are examined in their work. The results reveal that varied infill patterns have a considerable influence on the mechanical characteristics of 3D constructed products. In comparison to the other infill patterns studied, the rectilinear pattern performed best, with a significant modulus of elasticity, tensile strength, and good resistance of impact (Cabreira & Santana 2020).

It was also the most cost-effective and mass-produced of the four patterns, using less material and taking less time to print while maintaining good mechanical characteristics. The mechanical behaviour of the triangular pattern was found to be much poorer than that of the other patterns studied. This is due to the interior structure's fragility. Honeycomb pattern had the longest processing time and heaviest sample mass, as well as a higher braking strain. To summarise, the infill pattern should be selected not only for mechanical performance but also for printing duration and cost of materials.

The capacity density of the material that will be deposited inside the item is called infill density. A lesser amount of infill necessitates less time of

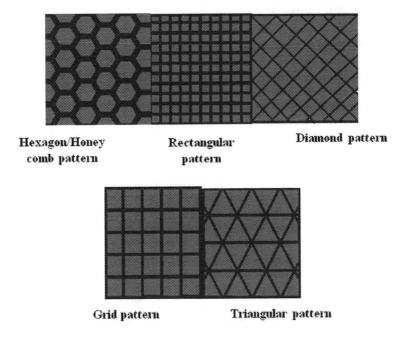

Figure 2.7 Various types of infill pattern.

build, nevertheless the mechanical characteristics of the printed part suffer, whereas dense infill necessitates significantly more time of build, but the mechanical characteristics of the produced part increase. The experiments on PLA and the results of the experimental testing on FDM specimens revealed that the infill % and the construction orientation have the biggest effect on the structural stability (Cerda-Avila et al. 2020).

The effect of FDM process factors on mechanical characteristics of 3D-printed carbon fibre-reinforced PLA composites is presented in this research. The experiment examines the consequence of percentage of infill, direction of build, and height of layer on the mechanical characteristics of the component on their own.

All three factors influenced the tensile property; however, in the case of % of infill and layer thickness, there is a number overhead whose effects are significantly stronger than at lower values.

When the thickness of layer is 0.25 mm and the percentage of infill is 80%, the greatest tensile strength is achieved. In the Izod test results, it is discovered that the thickness of layer has little effect on strength; however, building orientation and infill percentage had a significant impact. The goal of their research was to find the optimum alternative for 3D-printed objects that required impact or tensile strength from the supplied process parameter settings (Kamaal et al. 2021).

2.3.5 Build orientation

It explains how the given component is modified on the build base in relation to the machine tool's three principal axes, X, Y, and Z, as shown in Figure 2.8. When the sample's orientation to be built was changed from 0° to 90°, Feng et al. (2019) and Ashtankar et al. (2013) found a decrement in compressive and tensile strength.

The influence of orientation of build on the modulus of storage, loss modulus, and mechanical damping is identical to the angle of raster on these characteristics. Its impact on mechanical characteristics, on the other hand, is more obvious. The dynamic mechanical qualities deteriorate as the build orientation varies from 0° to 90°. This is owing to the detail that the quantity of layers needed to manufacture a vertically orientated item increases. As a result, skewed development converts a major factor, resulting in poor performance (Mohamed et al. 2016). The part accuracy of components manufactured by the FDM technique is investigated in this research. The effect of volumetric change and geometric inaccuracy of the "curl" kind on the parts was investigated (Gurrala & Regalla 2014). The roughness of the surface of FDM products is great when the component is inclined at the angles between 20° and 30° to the plate of build, according to the experiment (Pugalendhi 2012). Pandey et al. (2004) have used a genetic algorithm

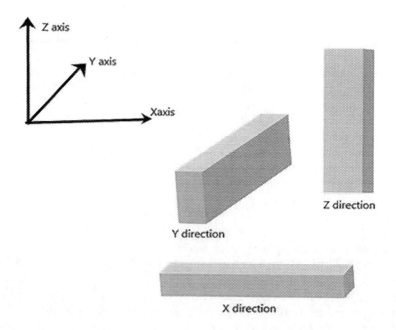

Figure 2.8 Schematic representation of the orientation of build.

to evaluate the optimum orientation in order to advance the finish of the surface and reduction in build time. The investigation revealed that the orientation angles of 0° and 90° were shown to be the most beneficial for good surface quality, build times, and budget (Rattanawong et al. 2001). The inclination of the model, build angles between 40° and 60° result in a lower surface polish and more cost because the maximum support material is utilised (Dani et al. 2013).

2.3.6 Air gap

It is defined as the space amid two neighbouring bead depositions. Figure 2.9 shows various types of air gap. When the air gap is found to be zero, the materials are in direct contact after deposition. If it is positive, it means the structure is loose, and if it is negative, it means the structure is dense (Solomon et al. 2021). Popescu et al. (2018) have concluded in their research that forming a negative bead-to-bead air gap improves mechanical characteristics.

The negative air gap enhanced the substance's flexural strength. This could be due to the compact structure produced by slightly touching filament deposition, which prevents gaps between fibres. This negative bead-to-bead air gap also improves the bond among end-to-end filaments, which improves the material's flexural characteristics. The negative air gap, on the other hand, has the disadvantage of lowering dimensional precision and quality of surface. Thus, the surface finish of such components having negative bead-to-bead air gap is clearly worse to that of their equivalents with a positive bead-to-bead air gap caused by filament overlapping and more material deposition (Gebisa & Lemu 2018).

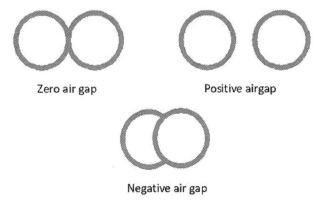

Figure 2.9 Three types of air gap.

2.3.7 Extrusion temperature

Temperature requirements for the FDM process are highly strict, and they are controlled by a control panel. Material bonding qualities, deposition properties, liquidity of the material filament, and width of the filament that is being extruded are all determined by temperature of extrusion. If this temperature at which the material is extruded is too less, then there will be an increase in the viscosity of material, the speed at which the material is being extruded will get reduced putting more strain on the extrusion system, and, in the worst-case scenario, clogging the nozzle, lowering adhesive strength in between layers of material, or even causing interparticle debonding.

If the temperature is too large, the coefficient of viscosity for the material decreases due to the priority for liquid, and the flowing capability increases. This results in filament material that is not exactly controlled in size, molecules in the preceding layer that tear, dissolve, or destroy before it is totally cooled due to pressure applied by the incoming layer, and surface roughness and warm undertones (Peng & Wang 2010).

2.3.8 Build time

When it comes to building a component, the amount of time it takes to construct it is critical. Because the time of build is mostly reliant on the height of a component, if same model is oriented differently, the time of build will vary. Furthermore, depending on the quantity of the support system, build times for SLA and FDM can vary significantly. Time required for data preparation, part construction time, and time required for post-processing are the three main components of build time.

When compared to component building and post-processing processes, preparation of data takes less time. It is difficult to compute an exact build time because all of the process variables, including nozzle acceleration and deceleration, must be taken into account. As a result, the overall build time in this study is approximated predicated on part geometry and machine characteristics. The estimated build time t_{build} is then given by Byun and Lee (2006):

$$t_{\text{build}} = t_{\text{ec}} + t_{\text{ia}} + t_s + t_m$$

t_{ec} is the time required for making the contour externally, t_{ia} is the time utilised for doing the interiors, t_s is the time needed for generating support, and t_m is the time that is dependent on the machine.

The most important process variable which affected the build duration was build orientation, according to them. A real coding GA was used to determine the best orientation of build based on the results. They compared the models' prediction skills to those of other articles published. It was

Parametric study of fused deposition modelling   33

obvious that the proposed model agreed with the preceding results in a reasonable way (Thrimurthulu et al. 2004).

The variables like thickness of layer and air gap could have a considerable impact on build time. The thickness of layer and air gap were also said to contribute 66.57% and 30.77% of the build time, respectively. The findings also suggested that 0.330 mm of layer height, 0.020 mm gap in air, and 30 raster angles were the best parameters for reducing construction time (Nancharaiah 2011). Table 2.2 shows the FDM studies available in literature.

Table 2.2 Some of the fused deposition modelling projects of various researchers

S. no	Project title	Input parameters	Output
1	"Optimization of rapid prototyping parameters for production of flexible ABS object" (Lee et al. 2005)	"Air gap, raster angle, raster width, layer thickness"	Elastic performance
2	"Improving dimensional accuracy of fused deposition modelling processed part using grey Taguchi method" (Sood et al. 2009)	"Part orientation, road width, layer thickness, air gap, raster angle"	Dimensional accuracy
3	"Investigation of the mechanical properties and porosity relationships in fused deposition modelling-fabricated porous structures" (Ang et al. 2006)	"Air gap, raster width, build orientation, build laydown pattern, build layer"	"Porosity, compressive yield strength, compressive modulus"
4	"Investigating the influence of infill percentage on the mechanical properties of fused deposition modelled ABS parts" (Alvarez et al. 2016)	"Infill percentage, raster angle, layer thickness, velocity temperature"	Mechanical properties
5	"A study on dimensional accuracy of fused deposition modeling (FDM) processed parts using fuzzy logic" (Sahu et al. 2013)	"Layer thickness, orientation, raster angle, raster width, air gap"	Dimensional accuracy
6	"The impact of process parameters on mechanical properties of parts fabricated in PLA with an open-source 3- D printer" (Lanzotti et al. 2015)	"Layer height, raster angle and number of shell perimeter"	"Tensile strength"
7	"Dynamic response of FDM made ABS parts in different part orientations" (Jami et al. 2013)	"Build orientations"	"High-strain-rate behavior"
8	"A study of the influence of process parameters on the mechanical properties of 3D printed ABS composite" (Christiyan et al. 2016)	Infill percentage, raster angle, layer thickness, velocity temperature	Maximum tensile and flexural strength

(Continued)

34 Additive Manufacturing

Table 2.2 (Continued) Some of the fused deposition modelling projects of various researchers

S. no	Project title	Input parameters	Output
9	"Fused deposition modelling (FDM) process parameter prediction and optimization using group method for data handling (GMDH) and differential evolution (DE)" (Rayegani & Onwubolu 2014)	"Part orientation, raster angle, raster width, air gap"	Tensile strength
10	"Effect of FDM process parameters on mechanical properties of 3D-printed carbon fibre–PLA composite" (Kamaal et al. 2021)	"Infill percentage, building direction and layer height"	Tensile and impact properties
11	"Studies on profile error and extruding aperture for the RP parts using the fused deposition modelling process" (Chang & Huang 2011)	Contour width, contour depth, raster width, raster angle	"Profile error and extruding aperture"
12	"A study on the influence of process parameters on the viscoelastic properties of ABS components manufactured by FDM process" (Dakshinamurthy & Gupta 2018)	"Raster angle, slice height, raster width"	Viscoelastic properties
13	"Evaluation of the influence of parameters of FDM technology on the selected mechanical properties of models" (Kozior & Kundera 2017)	Build orientations, raster angle	"Rheological properties"
14	"Optimizing process parameters of fused deposition modeling by taguchi method for the fabrication of lattice structures" (Dong et al. 2018)	"Print speed, nozzle temperature, layer thickness, fan speed"	"Print quality, elastic modulus and ultimate tensile strength"
15	"Mechanical properties of PLA graphene filament for FDM 3D printing" (Camargo et al. 2019)	Layer thickness, infill pattern	Mechanical properties

FDM finds its application in aircraft industry for making its various parts.

It is used in medical industry such as tissue engineering, human body part transplantation, paramedical products.

It is used to make parts in automobile industry.

It is used in textile industry.

It is used to make common household products.

It finds its application where products need asthetic details and environment friendly.

Finds application in automation industry.

Used to manufacture customised product.

It also find its application in food producing industry.

Figure 2.10 Applications of fused deposition modelling.

2.3.9 Materials in fused deposition modelling

The various types of materials that are being utilised in FDM are as follows (Lee et al. 2017, Unagolla & Jayasuriya 2020, Daminabo et al. 2020): ABS, M-30 ABS thermoplastic, ABSi thermoplastic, ABS-M30i thermoplastic PC(Polycarbonate)-ABS thermoplastic, ABS-ESD7 (electrostatic dissipative), PC thermoplastic (polycarbonate), PC-ISO thermoplastic, ULTEM 9085, PPSF/PPSU (polyphenylsulfone) thermoplastic, Nylon12, PLA, Hydrogel, and PEEK.

2.3.10 Application

Fused deposition modelling is an AM technology involving various materials extruding molten filament to create geometrically complicated shaped prototypes and parts. The various areas where the FDM finds its application are shown in Figure 2.10.

2.4 POST-PROCESSING TECHNIQUES

Types of post-processing techniques are shown in Figure 2.11. Initial techniques focused on mechanically slicing or compressing the top of surface profiles. The procedures are based on traditional metal finishing processes, although their performance on FDM materials differs significantly from that of metals. A chemical finishing method is promising for enhancing the quality of surface of parts, but more research is needed before they can be used on a large scale (Chohan & Singh 2017).

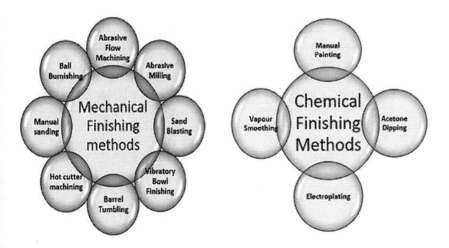

Figure 2.11 Post-processing techniques.

2.5 LIMITATIONS AND CHALLENGES

The primary issues with FDM technology today include a lack of in-depth research on associated FDM equipment, low manufacturing efficiency, poor accuracy, and excessive cost, among others. Although the adhesion between layers is tight, FDM technology's results are not as good as those cast and produced by a homogeneous conventional mould due to the inevitable accumulation processes. On the other hand, the moulding material's restrictions limit the scope of its applicability. Because of its potential to completely revolutionise industry, AM has been labelled as "fourth industrial revolution." Unfortunately, the aforementioned FDM revolution would require the ability to build any part in any geometry in only one manufacturing process requiring no or little worker expertise and no assembling, ending in a more cost-effective solution than the present production techniques.

The claim that AM has no geometric limits along with the existence of choice of design. Interior structures are a good illustration of geometries that are not viable to manufacture with CNC machining. However, this statement is only partially correct, because AM, particularly metal AM, lacks the ability to construct the so-called unsupported geometries like cube and solid sphere.

Another concern is that AM can produce a wide range of product quality. The most prevalent machines, namely, FDM 3D printers are less costly, whose producible parts' quality does not correspond to that of a high-end industrial AM system. Cost is a significant factor in the adoption of AM technology. The present cost of AM equipment is prohibitively expensive for both companies and individuals. AM technology is employed to create complex structures and the cost is substantially greater than products produced by large-scale manufacturing companies. Some individuals only observe AM parts manufactured with these less priced machines and believe they are representative of the AM technology's capabilities as a whole. These types of erroneous perceptions tend to hinder firms' adoption and exploitation of AM without causing structural damage.

Currently, it is difficult to say whether FDM will be more than a complement to the existing production methods. One can speculate that AM would revolutionise production in the future; there is some validity to this, but there is no technology on the horizon to address the previously described limitations of the AM process. Because only the infrastructure was missing in the 1990s, it was relatively straightforward to forecast the Internet's future importance. The technology was already in place, but it was too expensive.

Furthermore, equipment research and development for the fused deposition AM process is insufficient, and the accompanying industry needs further expansion. It has not yet reached the stage of large-scale industrial application.

2.6 CONCLUSION AND FUTURE SCOPE

When utilising a higher extrusion, wider diameter of nozzle, infill velocity, and more layer height, strength in tension and surface roughness increase greatly. The build time is reduced by increasing the velocity of filling the diameter of nozzle and the thickness of the layer.

Infill % and the construction orientation have the biggest effect on the structural stability. Negative air gap enhances the substance's flexural strength.

Print orientation is a critical factor that has an impact on the rheological characteristics of materials. Component orientation affects the anisotropic qualities of FDM parts, and it is critical for overall compressive properties. It is also safe to deduce, based on experimental results and general knowledge, that complicated infill form, more density of infill, and a large quantity of shells are desirable for enhancing compressive strength of the parts produced. According to the findings from the evaluated publications, it can be concluded that build time is the shortest when layer thickness is large, build orientation is zero, and infill density is low. The term "minimum build time" refers to how quickly an item can be printed.

Hybrid machines are being explored and developed that combine AM with traditional production processes. In this manner, the advantages of both procedures can be reaped while the downsides of each are mitigated. The combination of AM with biomaterials appears to be promising for purposes dealing with human health and wellness.

Currently, one of the most important topics in manufacturing is sustainability. Academics, administrations, and enterprises are concentrating their efforts on developing environmentally friendly production techniques. Efforts are being made to ensure that the unique AM processes are sustainable or, at the very least, have a low environmental impact. The main impetus for the adoption of AM technology is currently economic, with other reasons such as social and sustainability having little influence. Reliability, control, and quality of the parts must be ensured by AM, which necessitates the implementation of monitoring and inspection tools. Furthermore, research and development of equipment for the AM process is insufficient, and the accompanying industry needs further expansion.

REFERENCES

Ali, F., Chowdary, B. V., & Maharaj, J. (2014, September). Influence of some process parameters on build time, material consumption, and surface roughness of FDM processed parts: inferences based on the Taguchi design of experiments. In *Proceedings of the 2014 IACJ/ISAM Joint International Conference.*

38 Additive Manufacturing

Alvarez, K. L., Lagos, R., & Aizpun, M. (2016). Investigating the influence of infill percentage on the mechanical properties of fused deposition modelled ABS parts. *Ingeniería e Investigation*, 36(3), 110–116.

Ang, K. C., Leong, K. F., Chua, C. K. et al. (2006) Investigation of the mechanical properties and porosity relationships in fused deposition modelling-fabricated porous structures. *Rapid Prototyping Journal*, 12(2), 100–105.

Arumaikkannu, G., Uma Maheshwaraa, N., & Gowri, S. (2005). A genetic algorithm with design of experiments approach to predict the optimal process parameters for FDM. In *2005 International Solid Freeform Fabrication Symposium*.

Ashtankar, K. M., Kuthe, A. M., & Rathour, B. S. (2013, November). Effect of build orientation on mechanical properties of rapid prototyping (fused deposition modelling) made acrylonitrile butadiene styrene (abs) parts. In *ASME International Mechanical Engineering Congress and Exposition* (Vol. 56406, p. V011T06A017). American Society of Mechanical Engineers.

Byun, H. S., & Lee, K. H. (2006). Determination of the optimal build direction for different rapid prototyping processes using multi-criterion decision making. *Robotics and Computer-Integrated Manufacturing*, 22(1), 69–80.

Cabreira, V., & Santana, R. M. C. (2020). Effect of infill pattern in Fused Filament Fabrication (FFF) 3D Printing on materials performance. *Matéria (Rio de Janeiro)*, 25.

Camargo, J. C., Machado, A. R., Almeida, E. C., & Sousa Silva, E. F. M. (2019). Mechanical properties of PLA graphene filament for FDM 3D printing. *The International Journal of Advanced Manufacturing Technology*, 103(5/8), 2423–2443.

Cerda-Avila, S. N., Medellín-Castillo, H. I., & Lim, T. (2020). An experimental methodology to analyse the structural behaviour of FDM parts with variable process parameters. *Rapid Prototyping Journal*.

Chang, D. Y., & Huang, B. H. (2011). Studies on profile error and extruding aperture for the RP parts using the fused deposition modeling process. *The International Journal of Advanced Manufacturing Technology*, 53(9/12), 1027–1037.

Chaunier, L., Guessasma, S., Belhabib, S., Della Valle, G., Lourdin, D., & Leroy, E. (2018). Material extrusion of plant biopolymers: Opportunities & challenges for 3D printing. *Additive Manufacturing*, 21(March), 220–233. https://doi.org/10.1016/j.addma.2018.03.016.

Chohan, J. S., & Singh, R. (2017). Pre and post processing techniques to improve surface characteristics of FDM parts: A state of art review and future applications. *Rapid Prototyping Journal*.

Christiyan, K. G., Chandrasekhar, U., & Venkateswarlu, K. (2016). A study of the influence of process parameters on the mechanical properties of 3D printed ABS composite. *IOP Conference Series: Materials Science and Engineering*, 114, 1–8.

Dakshinamurthy, D., & Gupta, S. (2018). A study on the influence of process parameters on the viscoelastic properties of ABS components manufactured by FDM process. *Journal of the Institution of Engineers (India): Series C*, 99(2), 133–138.

Dakshinamurthy, D., & Gupta, S. (2018). A study on the influence of process parameters on the viscoelastic properties of ABS components manufactured by FDM process. *Journal of the Institution of Engineers (India): Series C*, 99(2), 133–138.

Daminabo, S. C., Goel, S., Grammatikos, S. A., Nezhad, H. Y., & Thakur, V. K. (2020). Fused deposition modeling-based additive manufacturing (3D printing): Techniques for polymer material systems. *Materials Today Chemistry*, 16, 100248. https://doi.org/10.1016/j.mtchem.2020.100248.

Dani, T. V., Kamdi, P. M., Nalamwar, G. C., & Borse, V. N. (2013). Multi objective optimization of built orientation for rapid prototyping of connecting rod. *International Journal of Scientific Research and Management*, 1(1), 13–18.

Dave, H. K., Patadiya, N. H., Prajapati, A. R., & Rajpurohit, S. R. (2021). Effect of infill pattern and infill density at varying part orientation on tensile properties of fused deposition modeling-printed poly-lactic acid part. *Proceedings of the Institution of Mechanical Engineers, Part C: Journal of Mechanical Engineering Science*, 235(10), 1811–1827.

Dong, G., Wijaya, G., Tang, Y., & Zhao, Y. F. (2018). Optimizing process parameters of fused deposition modeling by taguchi method for the fabrication of lattice structures. *Additive Manufacturing*, 19, 62–72.

Feng, L., Wang, Y., & Wei, Q. (2019). PA12 powder recycled from SLS for FDM. *Polymers*, 11(4), 727.

Galantucci, L. M., Bodi, I., Kacani, J., & Lavecchia, F. (2015). Analysis of dimensional performance for a 3D open-source printer based on fused deposition modeling technique. *Procedia CIRP*, 28, 82–87.

Gebisa, A. W., & Lemu, H. G. (2018). Investigating effects of fused-deposition modeling (FDM) processing parameters on flexural properties of ULTEM 9085 using designed experiment. *Materials*, 11(4), 500.

Griffiths, C. A., Howarth, J., Rowbotham, G. de-Almeida, & Rees, A. (2016). Effect of build parameters on processing efficiency and material performance in fused deposition modelling. *Procedia CIRP*, 49, 28–32. doi: 10.1016/j.procir.2015.07.024.

Gurrala, P. K., & Regalla, S. P. (2014). DOE based parametric study of volumetric change of FDM parts. *Procedia Materials Science*, 6, 354–360.

Haghighi, A., & Li, L. (2018). Study of the relationship between dimensional performance and manufacturing cost in fused deposition modeling. *Rapid Prototyping Journal*.

Hegab, H. A. (2016). Design for additive manufacturing of composite materials and potential alloys: A review. *Manufacturing Review*, 3, 11. https://doi.org/10.1051/mfreview/2016010.

Jami, H., Masood, S. H., & Song, W. Q. (2013). Dynamic response of FDM made ABS parts in different part orientations. *Advanced Materials Research*, 748, 291–294.

Kamaal, M., Anas, M., Rastogi, H., Bhardwaj, N., & Rahaman, A. (2021). Effect of FDM process parameters on mechanical properties of 3D-printed carbon fibre–PLA composite. *Progress in Additive Manufacturing*, 6(1), 63–69.

Journalof Composite Materials, 10(03), 45.

Kozior, T., & Kundera, C. (2017). Evaluation of the influence of parameters of FDM technology on the selectedmechanical properties of models. *Procedia Engineering*, 192, 463–468.

Lanzotti, A., Grasso, M., Staiano, G., & Martorelli, M. (2015). The impact of process parameters on mechanical properties of parts fabricated in PLA with an open-source 3- D printer. *Rapid Prototyping Journal*, 21(5), 604–617.

40 Additive Manufacturing

Lee, B. H., Abdullah, J., & Khan, Z. A. (2005). Optimization of rapid prototyping parameters for production of flexible ABS object. *Journal of Materials Processing Technology*, 169(1), 54–61.

Lee, J. Y., An, J., & Chua, C. K. (2017). Fundamentals and applications of 3D printing for novel materials. *Applied Materials Today*, 7, 120–133. https://doi.org/10.1016/j.apmt.2017.02.004.

Mohamed, O. A., Masood, S. H., Bhowmik, J. L., Nikzad, M., & Azadmanjiri, J. (2016). Effect of process parameters on dynamic mechanical performance of FDM PC/ABS printed parts through design of experiment. *Journal of Materials Engineering and Performance*, 25(7), 2922–2935.

Nancharaiah, T. (2011). Optimization of process parameters in FDM process using design of experiments. *International Journal on Emerging Technologies*, 2(1), 100–102.

Nancharaiah, T. R. D. R. V., Raju, D. R., & Raju, V. R. (2010). An experimental investigation on surface quality and dimensional accuracy of FDM components. *International Journal on Emerging Technologies*, 1(2), 106–111.

Novak-Marcincin, J., Novakova-Marcincinova, L., Barna, J., & Janak, M. (2012). Application of FDM rapid prototyping technology in experimental gearbox development process. *Tehničkivjesnik*, 19(3), 689–694.

Pandey, P. M., Thrimurthulu, K., & Reddy, N. V. (2004) Optimal part deposition orientation in FDM by using a multicriteria genetic algorithm. *International Journal of Production Research*, 42(19), 4069–4089.

Peng, A. H., & Wang, Z. M. (2010). Researches into influence of process parameters on FDM parts precision. *Applied Mechanics and Materials*, 34–35, 338–343.

Popescu, D., Zapciu, A., Amza, C., Baciu, F., & Marinescu, R. (2018). FDM process parameters influence over the mechanical properties of polymer specimens: A review. *Polymer Testing*, 69, 157–166.

Pugalendhi, S. (2012). Experimental investigation of surface roughness for fused deposition modelled part with different angular orientation. *ADMT Journal*, 5(3).

Qattawi, A., Alrawi, B., & Guzman, A. (2017). Experimental optimization of fused deposition modelling processing parameters: A design-for-manufacturing approach. *Procedia Manufacturing*, 10, 791–803.

Rajpurohit, S. R., & Dave, H. K. (2018). Effect of process parameters on tensile strength of FDM printed PLA part. *Rapid Prototyping Journal*.

Rattanawong, W., Masood, S. H., & Iovenitti, P. (2001). A volumetric approach to part build orientations in rapid prototyping. *Journal of Materials Processing Technology*, 119, 348–353.

Rayegani, F., & Onwubolu, G. C. (2014). Fused deposition modelling (FDM) process parameter prediction and optimization using group method for data handling (GMDH) and differential evolution (DE). *The International Journal of Advanced Manufacturing Technology*, 73(1–4), 509–519.

Sahu, R. K., Mahapatra, S., & Sood, A. K. (2013). A study on dimensional accuracy of fused deposition modeling (FDM) processed parts using fuzzy logic. *Journal for Manufacturing Science and Production*, 13(3), 183–197.

Solomon, I. J., Sevvel, P., & Gunasekaran, J. (2021). A review on the various processing parameters in FDM. *Materials Today: Proceedings*, 37, 509–514.

Sood, A. K., Ohdar, R., & Mahapatra, S. (2009). Improving dimensional accuracy of fused deposition modelling processed part using grey Taguchi method. *Materials & Design*, 30(10), 4243–4252.

Standard, A. S. T. M. (2013). F2792-12a: Standard terminology for additive manufacturing technologies (ASTM International, West Conshohocken, PA, 2012). *Procedia Eng*, 63, 4–11.

Taylor, G., Wang, X., Mason, L., Leu, M. C., Chandrashekhara, K., Schniepp, T., & Jones, R. (2018). Flexural behavior of additively manufactured Ultem 1010: Experiment and simulation. *Rapid Prototyping Journal*.

Thrimurthulu, K., Pandey, P. M., & Reddy, N. V. (2004). Optimum part deposition orientation in fused deposition modeling. *International Journal of Machine Tools and Manufacture*, 44(6), 585–594.

Unagolla, J. M., & Jayasuriya, A. C. (2020). Hydrogel-based 3D bioprinting: A comprehensive review on cell-laden hydrogels, bioink formulations, and future perspectives. *Applied Materials Today*, 18, 100479. https://doi.org/10.1016/j.apmt.2019.100479.

Vasudevarao, B., Natarajan, D. P., Henderson, M., & Razdan, A. (2000, August). Sensitivity of RP surface finish to process parameter variation. In *Solid Freeform Fabrication Proceedings* (pp. 251–258). Austin, TX: The University of Texas.

Vijay, P., Danaiah, P., & Rajesh, K. V. D. (2011). Critical parameters effecting the rapid prototyping surface finish. *Journal of Mechanical Engineering and Automation*, 1(1), 17–20.

Wohlers, T., & Gornet, T. (2014). History of additive manufacturing. *Wohlers Report*, 24, 118.

Wu, W., Geng, P., Li, G., Zhao, D., Zhang, H., & Zhao, J. (2015). Influence of layer thickness and raster angle on the mechanical properties of 3D-printed PEEK and a comparative mechanical study between PEEK and ABS. *Materials*, 8(9), 5834–5846. doi: 10.3390/ma8095271.

Yang, L., Li, S., Li, Y., Yang, M., & Yuan, Q. (2019). Experimental investigations for optimizing the extrusion parameters on FDM PLA printed parts. *Journal of Materials Engineering and Performance*, 28(1), 169–182.

Chapter 3

Microstructural and mechanical properties of aluminium metal matrix composites developed by additive manufacturing—A review

Amarish Kumar Shukla and J. Dutta Majumdar
Indian Institute of Technology (Kharagpur)

CONTENTS

3.1 Introduction	43
3.2 Classifications and fabrication techniques	45
3.2.1 Powder bed fusion process	46
3.2.1.1 Selective laser melting	47
3.2.1.2 Selective laser sintering	47
3.2.1.3 Electron beam melting	48
3.3 Effect of process parameter on microstructure and mechanical properties	49
3.3.1 Microstructure	49
3.3.2 Mechanical properties	52
3.4 Advantages and limitations of additive manufacturing	53
3.5 Applications	54
3.6 Summary and future scope	54
References	55

3.1 INTRODUCTION

In conventional manufacturing, polymer is the generally used material by Additive Manufacturing (AM) process for industrial applications. But with time, aluminium (Al) is used as a base material for AM because of its availability, higher strength to weight ratio, and corrosion resistance for structural, automobile, and aerospace applications (Olakanmi, Cochrane, and Dalgarno 2015). However, monolithic Al alloys have lower strength, stiffness, and wear resistance, and as a result, the reinforcement of second ceramic particles into monolithic Al alloy requires tailoring the properties of matrix material to enhance its application on a large scale (Shukla and Dutta Majumdar 2019 a, b). The metal matrix composites (MMCs)

DOI: 10.1201/9781003258391-3

have combined properties of metals and ceramics such as toughness, ductility, high strength, high stiffness-to-weight ratio, high modulus, good wear resistance, outstanding chemical inertness, and the ability to withstand greater service temperatures, compared to epoxy or phenolic matrices (Dahotre, McCay, and McCay 1989). Due to these qualities of Al, MMCs have recently attracted a lot of attention (Hu et al. 2018; Li et al. 2021; Shukla, Mondal, and Dutta Majumdar 2021). In addition, the reinforcement particles improved the interactions between the laser beam and the powder particles, resulting in improving laser absorptivity and assisting in the resolution of the high laser reflectivity problem compared to monolithic Al alloy. SiC, Al_2O_3, TiC, TiB_2, B_4C, TiN, etc. are mostly used as reinforcements for the development of MMCs for commercial applications (Miracle 2005). Among these, SiC is the most common reinforcement used for commercial applications followed by Al_2O_3 and TiC (Li et al. 2021; Miracle 2005). Metallic composites may be developed by several techniques such as melting casting (M&C) and powder metallurgy (PM) process (Shukla and Dutta Majumdar 2021). However, these fabrication techniques have challenges such as segregation of reinforced particles, weak interfacial bonding, non-uniform distribution, and difficulty in machining these reinforced particles in casting routes (Dadkhah et al. 2021). On the other hand, composites developed by the PM process are prone to higher porosity and inferior mechanical properties, resulting in premature component failure (Shukla and Dutta Majumdar 2021). Besides, additive manufacturing (AM) provides uniform distribution of the secondary phase in the metallic matrix, as well as defect-free, dense, and high-strength components (Dadkhah et al. 2021). AM processes have a greater ability to fabricate complicated shaped components and offer advantages such as high design freedom, design optimisation, and on-demand production of customised parts compared to conventional manufacturing methods. AM is the method of combining materials to form objects through layer-by-layer deposition of material from specified 3D model data, and it is the process of creating the final shape by adding materials. This process does not require cutting tools, coolants, fixtures, and other auxiliary resources for the assembly (Huang et al. 2013). The AM process consists of three basic steps: (i) creating a computerised 3D solid model and converting it to a standard AM file format, such as the traditional STL (standard tessellation language) format; (ii) sending the file to an AM machine and manipulating it, such as changing the part's position and orientation or scaling; and (iii) building the part layer by layer on the AM machine (Huang et al. 2013). AM techniques have been widely adopted as a new manufacturing concept in a range of industries, including aerospace, automotive prototyping, and most recently, construction (Dadkhah et al. 2021; Han and Jiao 2019). SiC and Al_2O_3 reinforced

Al composites are commonly utilised in brake drums and cylinder liners for automotive applications, as well as rotor vanes and plates for structural aerospace parts (Manfredi et al. 2014; Olakanmi, Cochrane, and Dalgarno 2015). There are several literatures on composites developed using AM techniques, but there are only a few articles that focus on the fabrication routes, formation mechanisms, and effect of process parameters like powder morphology, volume per cent of reinforcement, scan speed, and power on the microstructure and mechanical properties of Al composites which need to be explained in detail.

In the present work, the methods and challenges associated with the fabrication of Al composite by AM processing will be addressed. The AM technologies are first introduced to provide a fundamental grasp of the AM methodology. Following that, various fabrication processes, microstructure, and mechanical properties are discussed. The final section discusses the advantages, limits, and future prospects of additively made MMCs. Finally, the key challenges in using metal AM methods to fabricate MMCs are examined and highlighted.

3.2 CLASSIFICATIONS AND FABRICATION TECHNIQUES

The AM process is classified into seven techniques, i.e., Vat photo polymerisation, binder jetting (BJ), sheet lamination (SL), material jetting (MJ), material extrusion (ME), powder bed fusion (PBF), and directed energy deposition (DED), based on the adhesion and bonding method (Dadkhah et al. 2021). These techniques are classified in terms of direct and indirect processes. Direct techniques are those that shape and consolidate components at the same time, eliminating the requirement for post-processing to densify the parts. In fact, the final components are created using these procedures by complete melting of metal powder and then solidifying it. Direct AM approaches include the laser-based PBF and DED procedures. While SL, BJ, MJ, and other indirect methods use a binder for shaping, some post-processing, such as debinding and sintering, is required to consolidate the materials and improve the density of the finished components. Wire or metal powder is utilised as the starting material in all metal AM methods. Using a heat source such as a laser or an electron beam, these powders are solidified into dense 3D shapes. The present work gives an overview of MMCs development using AM process, primarily laser-based PBF techniques. Selective laser melting (SLM), selective laser sintering (SLS), and electron beam melting (EBM) are three types of PBF technologies that are used to melt metal, ceramic, and polymeric material (Dadkhah et al. 2021).

3.2.1 Powder bed fusion process

The schematic and detailed operating principles of AM techniques are shown in Figure 3.1. A layer of powder is spread on the building platform or already solidified layers in PBF processes, and then melted selectively according to the CAD model. The process of spreading a layer of powder followed by selective melting by local heating is repeated until the entire portion is printed. PBF techniques are classified into two categories, i.e., laser powder bed fusion (LPBF), also known as selective laser melting (SLM), and electron beam melting (EBM). The focused source of energy used for the PBF technique can be either a laser beam (for SLM or SLS) or an electron beam (for EBM) process. Both PBF and EBM processes have similar concepts; however, their processing steps are different (Dadkhah et al. 2021; Azam et al. 2018). The powder particles are completely melted during the laser-based PBF process, resulting in significantly stronger final 3D components with a density of around 100% (Behera, Dougherty, and Singamneni 2019). In this process, a powder delivery system is used to spread a thin layer of metal powder across the metal build plate. The schematic of laser-based powder bed fusion system is shown in Figure 3.1. The powder particles melt and consolidate into a homogenous and dense component after a high-energy laser beam selectively scans the cross section of metal powder. Extra materials not included in the 3D model are employed as a support framework and are unaffected. After that, the build platform is lowered to a predetermined layer thickness, and the coating procedure for the next layer begins. After that, the next layer is laser scanned, and the process is continued until the final 3D component is created. When the part is finished, any

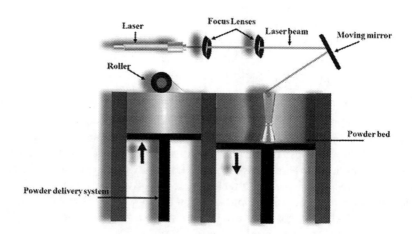

Figure 3.1 Schematic of laser-based powder bed fusion system.

leftover powders are collected and recycled directly from the construction chamber (Bhavar et al. 2017). The PBF process is further divided into a laser beam (SLM and SLS) and electron beam process which is explained in the next section.

3.2.1.1 Selective laser melting

SLM has been considered one of the promising powder bed AM techniques for metals and their matrix composites (Katz-demyanetz et al. 2019; Wang et al. 2020). It is a layer-by-layer AM process that can create complicated and high-performance components (Wang et al. 2020). In this process, a 3D CAD model is first created using an SLM device, then a thin layer of powder is applied to the base plate using a recoater, and finally, a high intensity laser energy source is utilised to fully melt and fuse the targeted regions. The base plate is lowered by a predetermined layer thickness after the melting and solidification of one layer, and the next powder layer is deposited on top of the preceding layer. To increase the surface quality and dimensional accuracy, layer-by-layer processing is utilised to manufacture the entire part with a difficult geometry. After a successful deposition of metal, an appropriate post-treatment such as mechanical polishing, electro-polishing, plasma polishing, and sandblast are used to improve the surface quality and dimensional accuracy of a component (Han and Jiao 2019). Al-Si-based Al alloys are commonly used materials to develop MMCs by AM process (Tang et al. 2021). As pure Al is extremely reflective of laser energies, only 7% of incident laser energy is absorbed while Si has very high absorptivity (about 70%). Because of this, Si-rich Al alloys (such as Al–12Si and Al–10Si–Mg) are the most used materials for laser-based AM processes (Tang et al. 2021).

3.2.1.2 Selective laser sintering

The SLS technique uses a high-powered laser to fuse small particles of the construction material. The powder bed is heated to slightly below the melting point of the material to minimise thermal distortion and allow fusing to the prior layer. Each layer is drawn on the powder bed with the laser to sinter the material. The sintered material forms the part, while the unsintered powder provides structural support and can be cleaned and recycled once the project is completed. In comparison to other AM technologies, SLS provides for the rapid creation of complicated parts that are more robust and functional. There is no need for postcuring, and the build time is minimal. This method, on the other hand, is challenging due to the numerous build variables that must be determined, as well as the difficulty of material changeover. Furthermore, the surface smoothness of the final components is inferior to that of SLM (Kamrani and Nasr 2010).

3.2.1.3 Electron beam melting

EBM is a PBF technique in which an electron beam is used to melt powder particles completely. Figure 3.2 shows a schematic of the EBM process. In this process, a high-power density electron beam source in a vacuum chamber is used to perform powder fusion. The vacuum environment (around 1×10^{-5} mbar) is used in this process to improve the electron beam quality, prohibition of the electron beam dissipation, and prevention of powder contamination of the fabricated components. In this method, electrons are emitted from a filament at a high temperature, and the electron beam is controlled by two magnetic fields moving at half the speed of light. One field acts as a magnetic lens, focusing the beam to the proper diameter, while another field deflects the concentrated beam, melting the desired locations. In this procedure, a layer of powder is dispersed over the build platform, followed by electron beam preheating and powder melting based on digital data. As stated in the CAD model, the build platform is lowered, and the next layer of powder is spread, followed by preheating and melted ageing. Finally, these steps are repeated to build the 3D components. As with other AM processes, the leftover powder should be removed from the build chamber once the finished part is created (Murr et al. 2012). When compared to traditional PBF methods, EBM is a faster AM method. This method favours the creation of structures with minimum residual

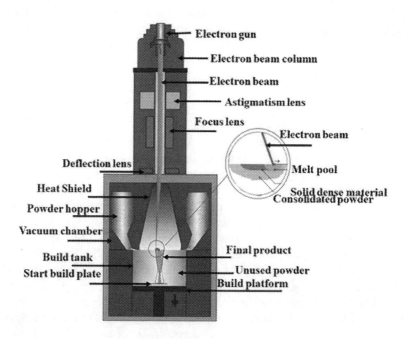

Figure 3.2 Schematic of electron beam melting.

stress. The vacuum procedure creates a completely clean and reaction-free environment. The thermal stress generated during the construction process is reduced by preheating the particles before melting. The reduction in residual stress reduces the danger of crack formation, allowing the manufacture of brittle materials like intermetallic. However, deposited layers are thicker (range of 45–150 μm) and the surface is rougher compared to the conventional PBF process, that is, 15–50 μm (Dadkhah et al. 2021).

3.3 EFFECT OF PROCESS PARAMETER ON MICROSTRUCTURE AND MECHANICAL PROPERTIES

The microstructure and mechanical properties of MMCs developed by laser PBF AM route depend on several parameters such as reinforcement particles, laser energy density, scanning speed, vol.% of reinforcement particle, powder morphology, scan speed, and power. These parameters are affected on a large scale, which is elaborated in the following section.

3.3.1 Microstructure

The absorptivity of laser energy depends on the addition of reinforcement particles that play a key role to tailor the microstructure of the developed composite by the laser PBF technique. Li et al. (2021) reported that the absorptivity of pure Al alloy (AlSi10Mg) by the PBF process varies in the range of 0.19–0.32, while the absorptivity of composite with the addition of secondary reinforced particles such as TiB_2 and SiC in AlSi10Mg matrix varied in the range of 0.34–0.49 (for TiB_2) and 0.37–0.59 (for SiC). The absorptivity of both the reinforced particles was significantly enhanced compared to AlSi10Mg powder. In addition, the results showed that SiC reinforcement was more effective than TiB_2 in improving laser absorption behaviours. As a result, it could be one of the causes of the vast range of SiC particles used in the PBF process. Furthermore, the incorporation of these reinforcing particles aids in tailoring the microstructure of the composite and refining the grain size of the developed composite (Li et al. 2021). Liu et al. reported that the addition of TiB_2 particles in laser-based PBF process shows a different crystallographic orientation and grain size. They reported that the Al alloy shows elongated grains while reinforcement of TiB_2 particles shows small and random grains, which is shown in Figure 3.3 (Tang et al. 2021; Xi et al. 2019). The size of reinforced particles affects the microstructure and mechanical properties of developed composite by using the AM route. In addition, it is reported that nanoparticles help to produce fine equiaxed grain growth and eliminate the hot cracking compared to pure Al7075 alloy which formed columnar growth of dendrite. However,

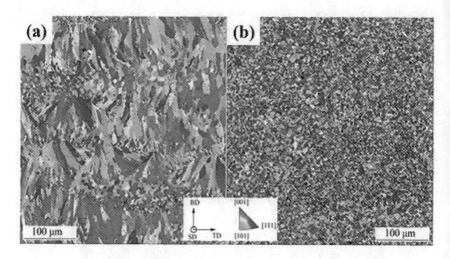

Figure 3.3 Electron Backscatter Diffraction (EBSD) inverse pole figure maps of (a) Al-12Si alloy and (b) Al-12Si/TiB^2 composite samples, and the colour indicates crystallographic orientation (Xi et al. 2019). (Permission granted by Elsevier, Licence Number-5347451291239.)

the micron size of reinforced particles of SiC in the Al matrix formed a crack at the interface of the developed composite. The mechanism of the formation of microstructure in Al and Al-MMCs is different. During the AM process, the Al powder is melted locally via laser irradiation, resulting in a melt pool that quickly solidifies via nucleation and directional growth mechanisms. Columnar microstructure is typical in monolithic Al alloys due to the lack of heterogeneous nucleation sites and the presence of a large temperature differential in the building direction (Chen et al. 2020; Wu et al. 2016). The Marangoni convection occurs in the melt pool when the laser interacts with the Al powder. It is a fluid flow mechanism that helps to ensure that reinforcing materials are distributed evenly throughout the liquid. However, because metal AM techniques use liquid material that solidifies quickly, there is not enough time for a negative interaction between the Al matrix and reinforcement phase. As a result of the quick solidification process, very fine second phases emerge, which, when combined with the unreacted reinforcement, form multiphase reinforced MMCs. The microstructure of MMCs changes from columnar to equiaxed with a small aspect ratio as a result of the presence of reinforcement and the fine second phase. It is also worth noting that, due to the large difference in the melting point difference between the Al matrix (about 660°C) and reinforcing materials (typically more than 2000°C), these reinforcements can greatly improve Al alloy laser absorption and process efficiency.

Microstructural and mechanical properties of aluminium metal   51

Dahotre, McCay and McCay (1989) developed MMCs by using a high-power (1–2.5 kW) CO_2 laser and the processing parameter is feed 0.045 cm beam diameter, transverse speed of 25 cm/s, and specific energy in the range of 5.3–13.0 J/m^2 to subsurface melt SiC particulate/A356-Al composites and to determine the feasibility of laser techniques for processing the MMCs. They reported that the fusion zone's microstructure varies depending on the specific energy. The variation in structure from a specific site in the fusion zone to a place in the unmelted region follows the same pattern for all laser treatments. In the fusion zone, plate-like and blocky precipitates are spread in a fine dendritic matrix, whereas in the surrounding region, plate-like precipitates, SiC particles, and small blocks of Si are detected. According to Pei Wang et al. (2020), grain structure variation is also influenced by scanning speed. The in situ and ex situ reactions have an impact on the microstructure of composites. Lijay et al. (2016) used an in situ reaction of inorganic salt K_2TiF_6 and ceramic particle SiC with molten Al to create AA6061/TiC MMCs. The findings revealed that a substantial number of TiC particles generated in situ were consistently distributed in intergranular regions, successfully refining the grains of the Al matrix. They reported that the reinforcement of TiC particles in metal matrix exhibits various shapes like spherical, cubic, and hexagonal, and the particles of TiC distributed uniformly with the clear interface and excellent bonding between the matrix and reinforcement. Because of that these particles of TiC improved the microhardness and mechanical properties of the MMCs. The MMCs generated by in situ reaction approach solve the shortcomings of particle surface pollution and poor reinforcement-matrix bonding as compared to MMCs formed by the direct adding of reinforcements. As a result, the in situ laser AM approach may be able to provide a superior overall performance.

The laser energy density also helps to tailor the microstructure of composite fabricated by AM routes. Lower the laser energy density, the coarser the microstructure; however, microstructure changed to fine by using a higher laser energy source. The influence of energy density on the microstructure development of the 7075 Al alloy by laser PBF processes was investigated by Li et al. (2021). They reported that when the energy density was low, the grains formed epitaxially, resulting in columnar grains with an average size of 4.1 μm. However, increasing the energy density caused the grains to transform from columnar to equiaxed, resulting in an increase in the number of equiaxed grains with an average size of 0.78 μm. According to Astfalck et al. (2017), the number of angular reinforced SiC particles decreased as the laser energy increases from 21 to 71 J/mm^3 as compared to a composite fabricated at lower laser energy. They also reported that the composite produced at increased energy density, a needle-like phase (Al_4C_3), developed along with large equiaxed Si particles. The microstructure of the composite revealed that the particle size of SiC reinforced decreases; however, a small amount of porosity is also formed in the matrix.

3.3.2 Mechanical properties

The distribution of reinforcement particles and the structure of the reinforcement-matrix interface determine the mechanical properties of MMCs developed by laser-based PBF process. The laser density and method used to introduce the reinforcement particles into the Al matrix and determine the interface characteristics of the composite, i.e., ex situ or in situ. The reinforcement of SiC in Al matrix by ex situ generates the Al_4C_3 phase, which is brittle, and forms cracks at the interface, reducing the ductility and tensile strength of the developed composite by using the laser-based PBF process. They also reported that cracks are also formed in Al_2O_3 reinforced composites. However, the formation of cracks can be avoided by increasing the laser energy density. The porosity also deteriorates the mechanical properties of the composite. The pores in the matrix are generated due to the addition of secondary reinforced ceramic particles which raises the viscosity of the melt and hence reduces flowability, resulting in porosity in the solidified portions. Pores and cracks are easily generated during the production of AlSi10MgTiC composites with insufficient laser energy (Gu et al. 2014; Tang et al. 2021). It is reported that when compared to micron-sized reinforcements, the reinforced nanoparticles improved the overall mechanical characteristics; however, reinforced micro range particles increase the hardness in comparison with Al alloys. In these systems, the microparticles act as stress concentration sites, hence their effect on mechanical characteristics is unstable. The flowability of nanoparticles is limited by their ease of agglomeration, which decreases the MMC powder mixture's processability. The $Al-Al_2O_3$ nanocomposites processed by laser PBF show a maximum 99.49% relative density which is obtained by optimising the process parameters by using the laser energy density of 317.5 J/mm^3 (Tang et al. 2021). The optimisation of process parameters also improved the tensile strength and ductility of the composite. They also reported that the relative density depends on the laser energy density and laser scanning speed which tailored the mechanical properties of the composite developed by the laser PBF process. Gu et al. (2014) reported that the tensile strength of AlSi10Mg-TiC nanocomposite increased from 400 to 452 MPa compared to the unreinforced AlSi10Mg alloy properties developed by the laser PBF process. The strength increased from 400 to 452 MPa, whereas the elongation of the composite showed similar behaviour. However, by increasing laser energy density, the tensile strength and elongation of the composite improved by 486 MPa and 10.9%, respectively. Li et al. developed the $AlSi10MgTiB_2$ (7 vol.%) nanocomposite by laser PBF process and studied the mechanical properties of this composite. They reported that the nanocomposite shows a high tensile strength (530 MPa), excellent ductility of (15.5%), and microhardness (191 HV). The possible strengthening mechanisms for composites developed by the laser-based PBF process are Hall-Petch strengthening, dislocation strengthening, load transfer effect, and Orowan strengthening (Tang et al. 2021). The current status and summary of the mechanical properties of AM processes are shown in Table 3.1.

Microstructural and mechanical properties of aluminium metal   53

Table.3.1 Current status and summary on the mechanical properties of the effect of different reinforcements on the mechanical properties by additive manufacturing processes

Authors	Matrix	Reinforcement and processing techniques	Mechanical properties
1	AlSi10Mg	SiC (8.5 vol.%), fabricated by the laser powder bed fusion (PBF) process technique	Hardness = 2.27 GPa Elastic modulus = 78.94 MPa
2	AlSi10Mg	SiC, fabricated by the laser PBF process technique	Relative density = 97.2% Microhardness = 218.5 HV Elastic modulus = 78.94 MPa
3	AlSi10Mg	TiN (4 wt.%), fabricated by the laser PBF process technique	Ultimate tensile strength = 491.2 MPa Yield strength = 315.4 MPa Elongation = 7.5%
4	Al-3.5-Cu-1.5-Mg-1Si	TiB_2 (5 vol.%), fabricated by the laser PBF process technique	Ultimate compressive strength = 500 MPa Yield compressive strength = 191 MPa
5	Al	Al_2O_3 (20 wt.%), fabricated by laser PBF process technique	Hardness = 148–175 HV
6	Al	Al_2O_3 (4 vol.%), fabricated by the laser PBF process technique	Hardness = 48 HV Ultimate tensile strength = 160 MPa Yield strength = 315.4 MPa Elongation = 5.1%
7	Al	TiC (17 vol.%), fabricated by the laser PBF process technique	Ultimate tensile strength = 500 MPa Yield strength = 300 ± 52 MPa Elongation = 3%
8	AlSi10Mg	TiB_2 (11.6 wt.%), fabricated by the laser PBF process technique	Hardness = 191 HV Ultimate tensile strength = 530 ± 16 MPa Yield strength = 315.4 MPa Elongation = 15.5%

3.4 ADVANTAGES AND LIMITATIONS OF ADDITIVE MANUFACTURING

AM has various advantages over other traditional techniques such as fast production rate, no requirement of jigs and fixtures, flexibility to develop complex products, and sustainability; however, rough final product, size limitations, cost, and the required skilled manpower may limit its use for a wide range of applications. The layer-by-layer manufacture of the specimen requires more time due to the high size of the sample. Furthermore, the surface morphology of the additively manufactured specimen is rough and ridged. As a result, it proceeds to the post-processing stage. The setup for AM

54 Additive Manufacturing

is thought to be expensive in terms of investment. Constant power supply is necessary to create the object; power consumption is also a major constraint. As a result, before AM can be employed in mass production, it must improve in terms of overall efficiency and process parameters control (Gu et al. 2012).

3.5 APPLICATIONS

Al composites developed by AM process possess uniform microstructures, higher hardness, and excellent mechanical properties. As a result, it is having several applications in components, for structural and functional applications. The important sectors where there is a need for Al-MMCs are automotive sectors, where the frame structure of automotive components, piston, vertical tailplane bracket, motorsports and aerospace interiors, rapid production of large aerospace components, engine breakthrough components and subsystems are mostly used. SiC-based Al composites are now widely employed in a variety of applications, from vehicle brake drums and cylinder liners to structural aerospace parts like rotor vanes and plates. Application of AM in the aerospace sector also includes aerospace interiors, rapid production of large aerospace components, engine breakthrough components and subsystems, including both military and commercial jets. Using SLM manufacturing technology, the team was able to design, construct, and test a vital rocket engine component. SLM was utilised to make parts for General Electric Aviation's forthcoming LEAP (Leading Edge Aviation Propulsion) turbofan engine series, which was developed in collaboration with Snecma of France (Jiao et al. 2018; Listani et al. 2017). There are several industries such as the energy sector, automobile, aerospace and aviation, transportation, and structural applications (Han and Jiao 2019). The energy conversion devices, electricity generation technology such as wind turbines and solar panels, and other conversion devices such as batteries and generators are examples of AM components.

3.6 SUMMARY AND FUTURE SCOPE

The composite developed by laser-based PBF process can fabricate complex components, have a uniform distribution of secondary reinforced particles, and improve the density, hardness, and mechanical properties compared with the monolithic Al alloys. On the other hand, a higher cooling rate generates residual stresses and defects of the AM process. In addition, agglomeration of powder particles, reinforcement of micron size particles, in situ or ex situ processes may increase the tendency of cracking at the interface and may create an obstacle to achieving excellent mechanical strength. The high processing costs, the demand for trained staff, and the manufacture of very particular components limit its application to a narrow range of applications.

As a result, more research, technological advancement, and material and method improvement are required to address some of these issues.

REFERENCES

Astfalck, L.C., G.K. Kelly, X. Li, and T.B. Sercombe. 2017. On the Breakdown of SiC during the Selective Laser Melting of Aluminum Matrix Composites. *Advanced Engineering Materials* 19(8): 1–6.

Azam, F.I., A.M.A. Rani, K. Altaf, T.V.V.L.N. Rao, and H.A. Zaharin. 2018. An In-Depth Review on Direct Additive Manufacturing of Metals. In: *IOP Conference Series: Materials Science and Engineering*: 1–8.

Behera, M.P., T. Dougherty, and S. Singamneni. 2019. Conventional and Additive Manufacturing with Metal Matrix Composites: A Perspective. *Procedia Manufacturing* 30: 159–166.

Bhavar, V. et al. 2017. A Review on Powder Bed Fusion Technology of Metal Additive Manufacturing. In: *Additive Manufacturing Handbook: Product Development for the Defense Industry* (September): 251–261.

Chen, B., X. Xi, C. Tan, and X. Song. 2020. Recent Progress in Laser Additive Manufacturing of Aluminum Matrix Composites. *Current Opinion in Chemical Engineering* 28: 28–35.

Dadkhah, M., M.H. Mosallanejad, L. Iuliano, and A. Saboori. 2021. A Comprehensive Overview on the Latest Progress in the Additive Manufacturing of Metal Matrix Composites: Potential, Challenges, and Feasible Solutions. *Acta Metallurgica Sinica (English Letters)* 34(9): 1173–1200.

Dahotre, N.B., T.D. McCay, and M.H. McCay. 1989. Laser Processing of a SiC/Al-Alloy Metal Matrix Composite. *Journal of Applied Physics* 65(12): 5072–5077.

Gu, D.D. et al. 2014. Selective Laser Melting Additive Manufacturing of TiC/AlSi10Mg Bulk-Form Nanocomposites with Tailored Microstructures and Properties. *Physics Procedia* 56(C): 108–116.

Gu, D.D., W. Meiners, K. Wissenbach, and R. Poprawe. 2012. Laser Additive Manufacturing of Metallic Components: Materials, Processes and Mechanisms. *International Materials Reviews* 57(3): 133–164.

Han, Q., and Y. Jiao. 2019. Effect of Heat Treatment and Laser Surface Remelting on AlSi10Mg Alloy Fabricated by Selective Laser Melting. *The International Journal of Advanced Manufacturing Technology* 102: 3315–3324.

Hu, Y. et al. 2018. Laser Deposition-Additive Manufacturing of TiB-Ti Composites with Novel Three-Dimensional Quasi-Continuous Network Microstructure: Effects on Strengthening and Toughening. *Composites Part B: Engineering* 133: 91–100.

Huang, S.H., P. Liu, A. Mokasdar, and L. Hou. 2013. Additive Manufacturing and Its Societal Impact: A Literature Review. *International Journal of Advanced Manufacturing Technology* 67(5–8): 1191–1203.

Jiao, L. et al. 2018. Femtosecond Laser Produced Hydrophobic Hierarchical Structures on Additive Manufacturing Parts. *Nanomaterials* 8(8).

Kamrani, A.K., and E.A. Nasr. 2010. *Engineering Design and Rapid Prototyping.* Springer.

Katz-demyanetz, A., V.V. Popov Jr., A. Kovalevsky, and D. Safranchik. 2019. Powder-Bed Additive Manufacturing for Aerospace Application: Techniques, Metallic and Metal/Ceramic Composite Materials and Trends. *Manufacturing Review 5.*

Li, N. et al. 2021. Laser Additive Manufacturing on Metal Matrix Composites: A Review. *Chinese Journal of Mechanical Engineering (English Edition)* 34(1).

Lijay, K.J., J.D.R. Selvam, I. Dinaharan, and S.J. Vijay. 2016. Microstructure and Mechanical Properties Characterization of AA6061/TiC Aluminum Matrix Composites Synthesized by in Situ Reaction of Silicon Carbide and Potassium Fluotitanate. *Transactions of Nonferrous Metals Society of China* 26(7): 1791–1800.

Listani, S. et al. 2017. Additive Manufacturing of 316L Stainless Steel by Electron Beam Melting for Nuclear Fusion Applications. *International Journal of Advanced Manufacturing Technology* 486(2): 234–245.

Manfredi, D. et al. 2014. Additive Manufacturing of Al Alloys and Aluminium Matrix Composites (AMCs). *Light Metal Alloys Applications* (August): 2–34.

Miracle, D.B. 2005. Metal Matrix Composites – From Science to Technological Significance. *Composites Science and Technology* 65(15–16 SPEC. ISS.): 2526–2540.

Murr, L.E. et al. 2012. Metal Fabrication by Additive Manufacturing Using Laser and Electron Beam Melting Technologies. *Journal of Materials Science & Technology* 28(1): 1–14.

Olakanmi, E.O., R.F. Cochrane, and K.W. Dalgarno. 2015. A Review on Selective Laser Sintering/Melting (SLS/SLM) of Aluminium Alloy Powders: Processing, Microstructure, and Properties. *Progress in Materials Science* 74: 401–477.

Shukla, A.K., and J. Dutta Majumdar. 2019a. Studies on Microstructure, and Mechanical, Properties of Aluminium Foam Prepared by Spray Forming Route. *Procedia Manufacturing* 35(March): 861–865.

Shukla, A.K., and J. Dutta Majumdar. 2019b. Studies on Wear Behavior of Aluminium Foam Developed by Spray Forming Route. *Materials Today: Proceedings* 19: 532–535.

Shukla, A.K., and J. Dutta Majumdar. 2021. Effect of Process Parameters on Microporosity and Nanomechanical Properties of Aluminium Cenosphere Composite Foam Developed by Spray Forming Route. *Journal of Thermal Spray and Engineering* 3(2): 74–80.

Shukla, A.K., D.P. Mondal, and J. Dutta Majumdar. 2021. Metallurgical Characteristics, Compressive Strength, and Chemical Degradation Behavior of Aluminum-Cenosphere Composite Foam Developed by Spray Forming Route. *Journal of Materials Engineering and Performance.*

Tang, S., R. Ummethala, C. Suryanarayana, J. Eckert, K.G. Prashanth, and Z. Wang. 2021. Additive Manufacturing of Aluminum Based Metal Matrix Composites—A Review. *Advanced Engineering Materials*: 1–17.

Wang, P. et al. 2020. 30 Transactions of Nonferrous Metals Society of China (English Edition) A Review of Particulate-Reinforced Aluminum Matrix Composites Fabricated by Selective Laser Melting.

Wu, J. et al. 2016. Microstructure and Strength of Selectively Laser Melted AlSi10Mg. *Acta Materialia* 117: 311–320.

Xi, L. et al. 2019. Effect of TiB2 Particles on Microstructure and Crystallographic Texture of Al-12Si Fabricated by Selective Laser Melting. *Journal of Alloys and Compounds* 786: 551–556.

Chapter 4

In situ process monitoring and control in metal additive manufacturing

Mukesh Chandra
BIT Sindri

Vimal K.E.K.
National Institute of Technology (Tiruchirappalli)

Sonu Rajak
National Institute of Technology (Patna)

CONTENTS

4.1	Introduction	57
	4.1.1 Background: In suit monitoring and control in additive manufacturing	58
4.2	Vision-sensing methods	61
	4.2.1 Inspections of powder deposition	62
	4.2.2 Melt pool observation	63
	4.2.3 Bead geometry inspection using optical imaging	64
4.3	Thermal sensing methods	65
	4.3.1 Temperature monitoring	65
	4.3.2 Defect detection	66
4.4	Acoustic sensing methods	66
	4.4.1 Observation and control of the parameters	67
	4.4.2 Defect detection using acoustic emission	68
4.5	Conclusion	70
	References	70

4.1 INTRODUCTION

After the introduction of the industrial revolution of 4.0 in the 19th century, the manufacturing machines became digitized and digitalized by using several digital twins such as the Internet of Things, cloud computing, robots and cognitive computing, thereby creating a smart factory of the future (Lasi et al., 2014). It gave birth to a smart manufacturing process called additive manufacturing (AM) (Dilberoglu et al., 2017). AM involves layer-by-layer deposition of materials using powder, sheet, or wire. The material that can be used for the fabrication of the

DOI: 10.1201/9781003258391-4

58 Additive Manufacturing

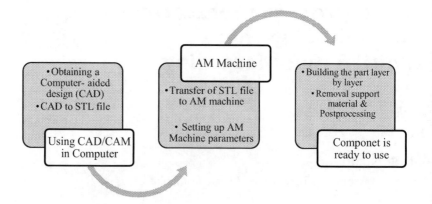

Figure 4.1 Flow diagram showing basic additive manufacturing steps to obtain a final product.

product varies from nonmetal, metal, alloys, composites to ceramics. The complex geometry of a miniature object can be fabricated using different AM techniques without the use of complicated tooling and fixture arrangement (Gibson et al., 2010a). Overall, the cost-saving on the final product as well as tooling and fixtures leads to the most economical fabrication process available these days. AM processes are known with different synonyms, i.e., rapid prototyping (RP), 3-D printing, stereolithography, additive fabrication, digital manufacturing, rapid manufacturing, solid freeform fabrication (SFF) and layered manufacturing (Patel, 2016; Vimal et al., 2021; Chandra et al., 2022). Figure 4.1 is a flow diagram showing basic AM steps to obtain a final product.

Metal additive manufacturing (MAM) has been an area of major interest in recent years to improve the process capability and productivity of manufacturing processes in various industries including core manufacturing for tool and die fabrication, shipbuilding, defence, aerospace, automotive, construction and medical as well (Zhang et al., 2018b; Blachowicz et al., 2021; Buchanan and Gardner, 2019; Salmi, 2021). MAM is classified into three categories, i.e., directed energy deposition (DED)-based AM, powder bed fusion (PBF)-based AM and sheet metal lamination (Zhang et al., 2018b). A detailed classification of MAMs along with their basic features is presented in Table 4.1. Several metals and alloys were utilized in the MAM for the fabrication of metallic parts (Bourell et al., 2017). Researchers have successfully fabricated Iron-based alloys (Guo et al., 2019), aluminium alloys (Horgar et al., 2018), titanium alloys (Kathryn et al., 2016), super nickel/chromium alloys, and other alloys using MAM.

4.1.1 Background: In suit monitoring and control in additive manufacturing

Major challenges are still faced by manufacturing industries to obtain a high level of quality and repeatability in MAM parts due to various factors such as the high complexity of the physical phenomena involved in

Monitoring and control in metal additive manufacturing 59

Table 4.1 Various additive manufacturing technologies for processing of metallic parts along with their important features

MAM	Subcategory	Features	Reference
Directed energy deposition (DED)	Direct metal deposition (DMD)	Laser beams are focused on the workpiece to produce melt pool, and metal powder is deposited to obtain metal deposition.	Dutta et al. (2011)
	Laser engineered net shaping (LENS™)	Laser beam produces a melt pool of metal on a substrate and metal powder is injected.	Atwood et al. (1998)
	Wire and arc additive manufacturing (WAAM)	An electric arc using a welding power source is used to produce a melt pool on the substrate and melt wire to obtain a high deposition volume of the metal layer by layer.	Pan et al. (2018)
Powder bed fusion (PBF)	Selective laser sintering (SLS)	It is the first commercialized PBF process. The powder is supplied on the substrate and heated using a laser beam layer by layer.	Kruth et al. (2003)
	Selective laser melting (SLM)	It used high-power solid-state laser and handles more fine powder compared to SLS under an inert atmosphere.	Leary et al. (2016)
	Direct metal laser sintering (DMLS)	High-power solid-state laser is exposed to metal powder in the liquid-phase sintering.	Khaing et al. (2001)
	Electron beam melting (EBM)	An electron beam is used to produce the fusion of metal powder and subsequent layers.	Galati and Iuliano (2018)
Sheet lamination	Laminated object manufacturing (LOM)	Ultrasonic or laser as an energy source to bond a metallic sheet.	Gibson et al. (2010b)

the fabrication process. The involvement of heat layer-by-layer results in residual stress and distortion in the MAM parts.

To overcome the challenges faced by manufacturing industries, much emphasis has recently been given to the "in situ process monitoring and control" for the MAM process. The first term "in situ process monitoring" refers to the ability of the manufacturing system to automatically detect various anomalies or defects during manufacturing operation, and later "control" refers to stopping or suppressing the defects. It can be achieved by feedback control with the help of various sensors. Artificial intelligence (AI) can help in the analysis of feedback in a more effective way for in situ process monitoring and control in the MAM process. A typical schematic diagram of in-suit process monitoring and control in MAM is shown in Figure 4.2. Nowadays, MAM systems are equipped with integrated sensing and monitoring tools to improve the performance of the MAM machines. A large number of MAM machines are being utilized these days for the fabrication of metallic

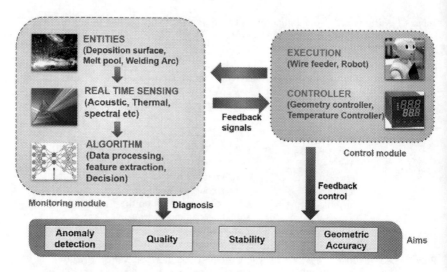

Figure 4.2 Flow diagram of entities extraction in metal additive manufacturing with in suit process monitoring and control.

components in industries. Newly set up industries may face the problem of finding suitable in suit monitoring devices for implication with AM machines.

In suit monitoring and sensing could help in performing multiple tasks during AM processes. They are as follows:

- In suit monitoring and sensing devices can help in detecting different anomalies in the AM process.
- It can send and receive feedback from sensors and help in the real-time correction of the problems.
- It can help in collecting a huge amount of data using different sensing devices for the purpose of process optimization in AM.

In recent years, much research has been carried out to understand the application and uses of in suit process monitoring and control in AM. Tapia and Elwany (2014) for the first time gave a comprehensive review of in suit process monitoring and control in MAM. Their review mainly consisted of an overview of some important sensing devices such as pyrometers, thermocouples and displacement sensors used in MAM. Everton et al. (2016) reviewed in situ monitoring systems separately for PBF and DED processes. Lu and Wong (2018) presented a review on monitoring and control systems using non-destructive testing techniques (NDT). Two NDT techniques, i.e., acoustic emission (AE) and thermal sensing, were reviewed in detail including their operating principles, various defects they can detect and their sensitivity. McCann et al. (2021) reviewed various in suit monitoring, including thermal, optical and acoustic sensing in PBF. Xia et al. (2020a) and Xu et al. (2018) reviewed the various process monitoring systems used in wire and arc additive manufacturing (WAAM), including acoustic signal, thermal signal, optical signal and X-ray. Later on, Chen et al. (2021) reviewed the different defect and detection approaches using in suit monitoring in WAAM. Traditional sensing tools are often used only for obtaining the exterior surface defects or anomalies in AM process and parts. In recent years, various sensing approaches have been applied for monitoring AM-fabricated parts like tomography, which can be used to produce three-dimensional images (3-D) of extracted features to analyse both exterior and interior portions (Plessis et al., 2018).

Various reviews were carried out for in suit process monitoring and control in AM; however, very few authors have presented a detailed review of in suit process monitoring and control in MAM. In suit monitoring and data acquisition both are important for improving the performance of MAM machines and processes. So, in this chapter, we have presented a detailed review of in suit process monitoring devices, their working principle and advantages and disadvantages. In addition, different data acquisition approaches, data processing and analysis are reviewed that could be helpful in newer design for additive manufacturing (DfAM), post-processing of MAM parts, optimization of MAM parameters and implementation of AI and machine learning (ML) for improving the MAM performances.

The rest of the chapter is organized as follows: Section 4.2 deals with in suit processing monitoring and control in MAM using vision sensing. Section 4.3 deals with the thermal sensing approach. Section 4.4 deals with the acoustic sensing approach. Section 4.5 concludes the chapter with a brief summarization of the sensing system used for in suit monitoring and control in MAM in a table.

4.2 VISION-SENSING METHODS

Vision-sensing methods for in suit process monitoring and control use optical and image sensors to obtain a position or image-based signal without any contact from the manufacturing system (Fang et al., 2010). Due to this non-contact feature, these sensors have got more attention in recent years. They are capable of obtaining high accuracy and an abundant amount of information flawlessly. Some of the important vision sensors used for manufacturing system monitoring in the past include charge-coupled device (CCD) camera (Xu et al. 2008), complementary metal oxide semiconductor (CMOS) camera (Fang et al., 2021), Circular laser 3-D scanner (Xu et al., 2008) and pyrometer and thermal cameras (Khanzadeh et al., 2017). Various image processing techniques were used for further processing of data to obtain useful information. The following section deals with various images' acquisition and processing techniques in detail.

4.2.1 Inspections of powder deposition

Powder-related process parameters in PBF such as distribution of powder, powder bed density and layer thickness are important for producing acceptable parts. Their monitoring during the process using traditional contact-based sensing devices is difficult. To solve the problem, an in situ process monitoring system for 3D topography measurement using enhanced phase measuring profilometry (EPMP) in PBF was proposed by Li et al. (2018). The authors used a novel slicing method of contour detection (Zhao et al., 2009) to improve the efficiency and accuracy. The camera and the projector were used to capture the fringe images of the powder layer. The images were processed using EPMP to develop dense 3D topography of the powder bed, which can help further in obtaining valuable information such as homogeneity, flatness, and defects. A schematic diagram of the proposed in suit monitoring system is shown in Figure 4.3. Craeghs et al. (2011) used a visual inspection consisting of a visual camera and three light sources for inspections of powder deposition in PBF. Illumination using a light source in three different directions, i.e., side, top and front of the build platform, was done to avoid any kind of shadow formation around the defects. The major defects that were detected in this study were local wear and local damage of the coater blade of the PBF system.

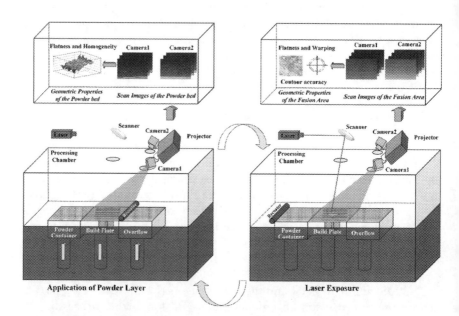

Figure 4.3 A schematic diagram showing a proposed model for in situ 3D monitoring of geometric signatures in the powder bed fusion process. (Adopted from Li et al., 2018.)

4.2.2 Melt pool observation

In situ monitoring of the melt pool region and melt pool temperature are the key technologies that can help in improving the part quality of MAM parts by predicting various anomalies during operation (Kwon et al., 2020). Real-time monitoring of melt pool to detect the process failure in selective laser melting (SLM) was carried out by Craeghs et al. (2012) using optical process monitoring. In this research, a mapping approach was used to obtain the melt pool data and thermal stresses due to deformation in a two-dimensional (2-D) plane. The mapping approach is a powerful tool to calculate and interpret the acquired data to obtain a mapping of the melt pool data on a 2-D plane. However, the authors did not propose any algorithms to obtain the important information from the mapping pictures. To improve the accuracy of in suit monitoring of melt pool further, Clijsters et al. (2014) designed an optical monitoring system integrated with a field-programmable gate array (FPGA) in an SLM machine control unit. In addition, the system consisted of two optical sensors, i.e., a planar photodiode sensor and a near-infrared thermal CMOS camera. Sensors helped in collecting high-quality images of the melt pool and FPGA helped in the processing of images at high speed opening the possibilities of improving the melt pool variation in the real-time process. Processed images on FPGA were able to estimate the melt pool–related parameters such as melt pool intensity, area, length and width. Moreover, the melt pool variations had an effect on pore formation. The scheme of the proposed optical measurement technique for online monitoring of melt pool is shown in Figure 4.4.

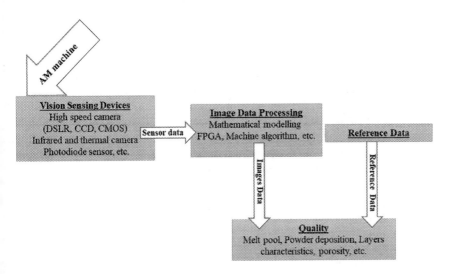

Figure 4.4 A schematic diagram of the optical measurement technique for online monitoring of melt pool.

In the most recent research, Khanzadeh et al. (2019) used thermal image monitoring of melt pool to predict the porosity in laser engineered net shaping (LENS)-fabricated components. A clustering technique, i.e., self-organizing maps (SOM), was used for image processing to obtain the morphological features and the temperature profile of the top surface of the melt pool. Le et al. (2021) used a high-speed camera to capture images of the melt pool in PBF. Some of the important parameters of camera setup include a frame rate of 7400 Hz, resolution of 128×128 pixels, a field of view (FOV) of 4×3.2 mm and the use of a 788–828 nm bandpass filter to remove the excess illumination due to laser beam. Captured images were processed using the following steps: (i) as the camera captured the side view of the melt pool, it was transformed to the top view using the mathematical perspective transformation method (Asselin et al., 2005), (ii) separation of the melt pool from the spatter using the image dilation process (Qidwai and Chen, 2009) and (iii) detection of the boundary of the melt pool using the intensity gradient approach (Luo and Shin, 2015). Later on, Fang et al. (2021) applied a U-net-based convolutional neural network (CNN) to obtain the morphology of the melt pool in SLM. The images were collected using a high-speed camera at 3000 frames per second of image size 1500×1988 pixels. In their case, the image processing involved the following steps: (i) extraction of a region of interest (ROI), (ii) segmentation of image using a U-Net, (iii) noise removal, and (iv) contour extraction.

Numerous other researches on melt pool monitoring to control the parameters in AM include QM Melt pool 3D monitoring system for LaserCUSING® process (Toeppel et al., 2016); a machine vision monitoring method consisting of support vector machines (SVM); CNN to monitor and extract the feature of the melt pool, plume and spatter in PBF (Zhang et al., 2018a); CNN-based porosity monitoring in LENS (Zhang et al., 2019); anomalies in WAAM using CNN (Lee et al., 2021); and prediction of the melt pool temperature distribution in LENS using long short-term memory (LSTM) and extreme gradient boosting (XGBoost) (Zhang et al., 2021).

4.2.3 Bead geometry inspection using optical imaging

The traditional method of bead geometry measurement in arc-based MAM using a stylus-based profilometer and coordinate measuring machine (CMM) could not be implemented for online in suit monitoring and control. A large amount of data acquisition and processing using these devices take too much time. Raw images captured using vision-based devices such as optical cameras contain a lot of information. Xiong and Zhang (2013) proposed a vision-sensing system consisting of two CCD cameras and composite filters for the detection of the bead height and width in WAAM. Xia et al. (2020b) developed a vision-based controlling of layer width in WAAM using a model predictive control (MPC) strategy. The width is estimated by

Monitoring and control in metal additive manufacturing **65**

extracting the edge of the welding pool captured using CCD. The image processing was carried out using an adaptive wiener filter (AWF) and the Canny algorithm. Later, Venturini et al. (2022) used a CMOS camera, a telephoto lens and an optical filter to acquire images of the deposited bead in WAAM. Compared to the CCD camera, the CMOS camera is cheaper and its setup can be easily reconfigured with software and hardware system as reported by the authors.

The vision-based monitoring system implemented with ML is getting a lot of popularity due to its large number of implications with AM machines for effective and accurate monitoring and control during the process.

4.3 THERMAL SENSING METHODS

A unique thermal cyclic behaviour is observed in the MAM process due to the involvement of heat in each deposited layer. Higher heat input can lead to deterioration of mechanical and metallurgical properties of AM-fabricated parts (Rosli et al., 2021). So, the in suit monitoring and control of temperature and heat are important during the AM process. Various contact and non-contact thermal sensors have been used for this purpose. A thermocouple is the most common type of contact-type thermal sensor used for monitoring temperature in almost all types of manufacturing operations. However, it has the limitation of a maximum temperature of 2000°C.

The introduction of an infrared (IR) camera made it possible for monitoring temperature in the highly complicated process of AM without making any contact. The IR camera works on the principle of thermography in which the thermal radiation (wavelength of 3–5 µm) range emitted by a source (i.e., deposited part of AM) gets converted into an electronic signal to produce a thermal image. Active and passive thermographies are two common techniques used for in suit process monitoring and control of various process parameters and anomalies in the AM process. In the passive technique, the natural heat distribution is enumerated over the surface of the target object to monitor the temperature. In the case of an active technique, induced the target item is subjected to induced heating or cooling in order to get a temperature profile across its surface.

4.3.1 Temperature monitoring

As discussed in Section 4.3, passive thermography techniques are generally used for measuring the temperature of the target object. Rodriguez et al. (2015) used an IR camera to acquire absolute temperatures of the melted or solid surfaces layer by layer in electron beam melting (EBM). Yang et al. (2017) used an IR camera to study the dimensions and volume of the molten pool and surface temperature of the deposited parts in Gas metal arc welding (GMAW)-based WAAM.

4.3.2 Defect detection

A pyrometer was used to collect thermal images of melt pool during fabrication of Ti-6Al-4V wall in laser-based additive manufacturing (LBAM) (Behnke et al., 2021). The data were collected in the form of comma-separated values (CSV) in a temperature range of 1000°C–2500°C at a frequency of 6.4 Hz. Each CSV file corresponds to one thermal image and contains a pixel matrix of size 752×480. Collected data were further used to develop ML models, i.e., model random forest classifier (RFC) and early stopping neural network (ESNN) to detect porosity in fabricated parts. Monitoring of undesired temperature gradients and porosity detection in PBF using IR thermography and computer tomography (CT) scanning techniques were carried out (Mireles et al., 2015). The result obtained using IR has substantially improved when compared with CT scanning techniques. Khanzadeh et al. (2017) captured melt pool thermal profile to predict porosity in Ti-6Al-4V thin walls fabricated using LBAM. Melt pool contains a large volume of raw data with a low signal-to-noise ratio and ill-structured data. So, data curing is important before processing it for analysis. A state-of-the-art data-driven modelling scheme consisting of spherical transformation, non-parametric curve fitting (bi-harmonic) method, SOM clustering and X-ray tomography characterization is applied to conclude the result. The experimental result revealed that the proposed model can predict the location of porosity almost 85% of the time.

Using ML algorithms for the processing of IR images can further improve the real-time process monitoring in AM. Baumgartl et al. (2020) used CNN for the processing of the thermographic image to detect and identify defects during printing in PBF. An ML model developed showed an accuracy of 96.80% without carrying any additional evaluations using X-ray or CT. A major limitation of IR camera is the limited penetration of light into the object to be detected. It can penetrate up to a depth of a few millimetres; hence, it is only suitable for detecting defects on the surface or upto a depth of a few millimetres (Avdelidis et al., 2011).

4.4 ACOUSTIC SENSING METHODS

An AE sensing is a class of NDT technique gaining a lot of popularity for in suit process monitoring and control in MAM-fabricated parts for detecting internal and surface defects and other anomalies (Hossain and Taheri, 2020). The inability of vision, and thermal sensors to penetrate metallic materials, the insufficiency of data acquisition due to improper sensor placement, and ineffective lighting cause certain challenges during in suit monitoring and control (Lu and Wong, 2018). As AE sensors work on the principle of propagation of acoustic wave and these waves

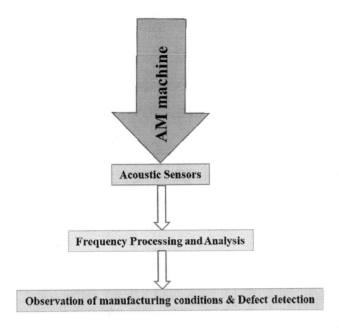

Figure 4.5 Workflow of acoustic emission sensing system used in additive manufacturing.

can propagate through or pass the surface, they do not have the problems mentioned above during data acquisition. The basic working flow of the AE sensing system is shown in Figure 4.5.

4.4.1 Observation and control of the parameters

Observation of process parameters allows for the real-time identification of major process failures during AM operation (Wu et al., 2016). The task becomes easier by collecting acoustic signatures such as noise signal using an AE sensor during operation. Taheri et al. (2019) used acoustic signatures for analysing the manufacturing operation in DED. Frequency responses collected were analysed using K-means statistical clustering algorithm. Plasma expansion of the arc in WAAM and interaction of the powder and the laser beam in laser-based DED are the main sources of AE (Hauser et al., 2022). Using an AE sensor consisting of an air acoustic microphone (PreSonus PRM1) and a welding camera (Cavitar C300), the authors investigated the geometrical fluctuations, track deviations, effect of shielding gas and nozzle-to-work distance during WAAM operation by observation of changing the intensity of the acoustic spectrum. Similarly, for laser-based DED track deviations, oxidation effects, process stability and other process parameters can be observed and controlled.

4.4.2 Defect detection using acoustic emission

AE sensors consisting of a piezoelectric element with an integral impedance converter have high sensitivity and a wide frequency range of 50–900 Hz. These sensors are capable of easily detecting plastic deformation of the material, cracks, pores, etc. in metallic materials in many manufacturing operations. Using the same type of AE sensor, Gaja and Liou (2018) detected different defects such as cracks and pores in a laser-based DED fabricated part. The frequency analysis was carried out with fast Fourier transformation, and further, an artificial neural network (ANN) and a logistic regression (LM) model were developed to detect the defects. Moreover, fibre bragg gratings (FBG) are capable of providing high sensitivity in signal detection because of the wide range of acoustic spectral range of operation (Hill and Meltz, 1997). Shevchik et al. (2018) used FBG optical fibre acoustic sensors for real-time monitoring of porosity in the fabricated part of SLM. The AE signal was processed and analysed using a ML approach, in particular, spectral CNN.

Another approach to capture high-frequency data of current and voltage in WAAM was developed in-house by Li et al. (2021). Their approach includes statistical feature identification, scoring and applying SVM ML techniques to detect a set of common defects. Ramalho et al. (2021) used acoustic sensing using a microphone to monitor and detect flaws in the WAAM-fabricated part. It is done by collecting the acoustic signal and measuring the arc voltage and the current during operation. The frequency distribution is analysed by calculating the power spectral density (PSD) and time-frequency using short time fourier transform (STFT).

Acoustic sensing techniques have shown tremendous growth in pursuit of improving the quality and capability of manufacturing various components because of their easy implementation for in suit processes monitoring and control since their discovery (Dornfeld, 1992). However, the major challenges with these sensing approaches are the unavailability of reliable acoustic devices, techniques for reduction of signal noises, accurate signal processing techniques, etc., which are still unexplored to a larger extent in the field of MAM.

Table 4.2 Summarization of various in suit process monitoring and control devices along with their application

Ref.	In suit monitoring devices	Measuring features	MAM method
Vision-sensing method			
Li et al. (2018)	Enhanced phase measuring profilometry (EPMP)	3D topography of the powder bed: flatness, homogeneity, defects of the powder bed	PBF

(Continued)

Monitoring and control in metal additive manufacturing   69

Table 4.2 (Continued) Summarization of various in suit process monitoring and control devices along with their application

Ref.	In suit monitoring devices	Measuring features	MAM method
Xiong and Zhang (2013)	CCD cameras with composite filter	Detection of the bead height and width	WAAM
Venturini et al. (2022)	CMOS camera, a telephoto lens with a filter	Image of the deposited bead	WAAM
Craeghs et al. (2011)	Camera with the light source	Detection of the defect in the coater blade of the PBF system	PBF
Le et al. (2021)	DSLR high-speed camera with filter	Melt pool	PBF
Fang et al. (2021)	Photodiode sensor and infrared thermal CMOS camera	Melt pool variations and pore intimation	SLM
Zhang et al. (2019), Khanzadeh et al. (2017, 2019)	Pyrometer and thermal camera	Prediction of porosity	LENS
Toeppel et al. (2016)	QM Meltpool 3D™ monitoring	Melt pool characteristics	LaserCUSING®
Zhang et al. (2018a)	Machine vision–based monitoring	Melt pool, plume and spatters	SLM
Thermal sensing method			
Behnke et al. (2021)	Pyrometer	Melt pool	LBAM
Mireles et al. (2015)	IR camera	Temperature gradients and porosity	PBF
Yang et al. (2017)	IR camera	The surface temperature of the deposited parts and molten pool	GMAW-based WAAM
Rodriguez et al. (2015)	IR camera	Absolute surface temperatures	EBM system
Acoustic sensing methods			
Ramalho et al. (2021)	Microphone-based acoustic sensing approach	To monitor and detect flaw production	WAAM
Koester et al. (2018)	AE system (Kistler 8152B111)	Detect cracks and pores	Laser-based DED
Hauser et al. (2022)	Acoustic emissions using air acoustic microphone	Geometrical fluctuations, track deviations, or oxidation effects	WAAM & laser-based DED

(*Continued*)

70 Additive Manufacturing

Table 4.2 (Continued) Summarization of various in suit process monitoring and control devices along with their application

Ref.	In suit monitoring devices	Measuring features	MAM method
Maier et al. (2011)	Acoustic emissions (AE) testing system	Internal structural defects	Laser-based DED
Gaja and Liou (2018)	Transducer (Digital Wave, B1025-MAE)	Cracks and delamination	Laser-based DED
Shevchik et al. (2019)	Fibre Bragg grating & machine leaning	Porosity	SLM (PBF)
Li et al. (2021)	In-house developed AE device	Set of common defects	WAAM

4.5 CONCLUSION

Table 4.2 summarizes the various in suit process monitoring and control sensors, i.e., vision, thermal and acoustics sensing systems used in MAM. Studies revealed that the main features of MAM, such as melt pool behaviour, deposited layer, temperature, the topography of the powder bed and various defects, are the main points of interest for in suit process monitoring and control in various research. The use of non-contact-based sensing systems such as vision systems, IR cameras and acoustics devices is gaining much popularity in the pursuit of improving the process capabilities of AM process and fabricated products. Moreover, the internal features and anomalies can be easily detected using signal-based acoustic sensors. Various ML algorithms used for processing the acquired data for real-time in suit process monitoring and control have shown improved accuracy compared with other traditional mathematical techniques. This review could help the industry as well as research and development organizations to understand the state-of-the-art of in suit process monitoring and control for MAM technologies.

REFERENCES

Asselin, M, E Toyserkani, M Iravani-Tabrizipour, and A Khajepour. 2005. Development of trinocular CCD-based optical detector for real-time monitoring of laser cladding. *IEEE International Conference Mechatronics and Automation* 3:1190–96.

Atwood, C, M Griffith, L Harwell, E Schlienger, M Ensz, J Smugeresky, T Romero, D Greene, and D Reckaway. 1998. Laser engineered net shaping (LENS): A tool for direct fabrication of metal parts. *International Congress on Applications of Lasers & Electro-Optics*:E1–7.

Avdelidis, NP, T-H Gan, C Ibarra-Castanedo, and XPV Maldague. 2011. Infrared thermography as a nondestructive tool for materials characterisation and assessment. *Thermosense: Thermal Infrared Applications XXXIII* 8013:308–14.

Baumgartl, H, J Tomas, R Buettner, and M Merkel. 2020. A deep learning-based model for defect detection in laser-powder bed fusion using in-situ thermographic monitoring. *Progress in Additive Manufacturing* 5(3):277–85. https://doi.org/10.1007/s40964-019-00108-3.

Behnke, M, S Guo, and W Guo. 2021. Comparison of early stopping neural network and random forest for in-situ quality prediction in laser based additive manufacturing. *Procedia Manufacturing*. https://doi.org/10.1016/j.promfg.2021.06.065.

Blachowicz, T, G Ehrmann, and A Ehrmann. 2021. Metal additive manufacturing for satellites and rockets. *Applied Sciences* 11(24):12036. https://doi.org/10.3390/app112412036.

Bourell, D, JP Kruth, M Leu, G Levy, D Rosen, AM Beese, and A Clare. 2017. Materials for additive manufacturing. *CIRP Annals – Manufacturing Technology* 66(2):659–81. https://doi.org/10.1016/j.cirp.2017.05.009.

Buchanan, C, and L Gardner. 2019. Metal 3D printing in construction: A review of methods, research, applications, opportunities and challenges. *Engineering Structures* 180(November 2018):332–48. https://doi.org/10.1016/j.engstruct.2018.11.045.

Chandra, M, F Shahab, KEK Vimal, and S Rajak. 2022. Selection for additive manufacturing using hybrid MCDM technique considering sustainable concepts. *Rapid Prototyping* 28 (7): 1297–1311.

Chen, X, F Kong, Y Fu, X Zhao, R Li, G Wang, and H Zhang. 2021. A review on wire-arc additive manufacturing: Typical defects, detection approaches, and multisensor data fusion-based model. *International Journal of Advanced Manufacturing Technology* 117(3–4):707–27. https://doi.org/10.1007/s00170-021-07807-8.

Clijsters, S, T Craeghs, S Buls, K Kempen, and JP Kruth. 2014. In situ quality control of the selective laser melting process using a high-speed, real-time melt pool monitoring system. *International Journal of Advanced Manufacturing Technology* 75(5–8):1089–101. https://doi.org/10.1007/s00170-014-6214–8.

Craeghs, T, S Clijsters, J-P Kruth, F Bechmann, and M-C Ebert. 2012. Detection of process failures in layerwise laser melting with optical process monitoring. *Physics Procedia* 39:753–59.

Craeghs, T, S Clijsters, E Yasa, and J-P Kruth. 2011. Online quality control of selective laser melting. In *2011 International Solid Freeform Fabrication Symposium*.

Dilberoglu, UM, B Gharehpapagh, U Yaman, and M Dolen. 2017. The role of additive manufacturing in the era of industry 4.0. *Procedia Manufacturing* 11:545–54.

Dornfeld, D. 1992. Application of acoustic emission techniques in manufacturing. *Ndt & E International* 25(6):259–69.

Dutta, B, S Palaniswamy, J Choi, LJ Song, and J Mazumder. 2011. Direct metal deposition. *Advanced Materials & Processes*:33.

Everton, SK, M Hirsch, PI Stavroulakis, RK Leach, and AT Clare. 2016. Review of in-situ process monitoring and in-situ metrology for metal additive manufacturing. *Materials and Design*. https://doi.org/10.1016/j.matdes.2016.01.099.

Fang, Q, Z Tan, H Li, S Shen, S Liu, C Song, X Zhou, Y Yang, and S Wen. 2021. In-situ capture of melt pool signature in selective laser melting using U-net-based convolutional neural network. *Journal of Manufacturing Processes* 68(PA):347–55. https://doi.org/10.1016/j.jmapro.2021.05.052.

Fang, Z, D Xu, and M Tan. 2010. Visual seam tracking system for butt weld of thin plate. *International Journal of Advanced Manufacturing Technology* 49(5–8):519–26. https://doi.org/10.1007/s00170-009-2421-0.

Gaja, H, and F Liou. 2018. Defect classification of laser metal deposition using logistic regression and artificial neural networks for pattern recognition. *The International Journal of Advanced Manufacturing Technology* 94(1):315–26.

Galati, M, and L Iuliano. 2018. A literature review of powder-based electron beam melting focusing on numerical simulations. *Additive Manufacturing* 19:1–20.

Gibson, I, D Rosen, and B Stucker. 2010a. Introduction and basic principles. In: *Additive Manufacturing Technologies*. Springer. https://doi.org/10.1007/978-1-4419-1120-9_1

Gibson, I, DW Rosen, and B Stucker. 2010b. Sheet lamination processes. In *Additive Manufacturing Technologies*, 223–52. Springer. https://doi.org/10.1007/978-1-4419-1120-9_8

Guo, Y, H Pan, L Ren, and G Quan. 2019. Microstructure and mechanical properties of wire arc additively manufactured AZ80M magnesium alloy. *Materials Letters*. https://doi.org/10.1016/j.matlet.2019.03.063.

Hauser, T, RT Reisch, T Kamps, AFH Kaplan, and J Volpp. 2022. Acoustic emissions in directed energy deposition processes. *The International Journal of Advanced Manufacturing Technology* 1(0123456789).

Hill, KO, and G Meltz. 1997. Fiber bragg grating technology fundamentals and overview. *Journal of Lightwave Technology* 15(8):1263–76.

Horgar, A, H Fostervoll, B Nyhus, X Ren, M Eriksson, and OM Akselsen. 2018. Additive manufacturing using WAAM with AA5183 wire. *Journal of Materials Processing Technology* 259(November 2017):68–74. https://doi.org/10.1016/j.jmatprotec.2018.04.014.

Hossain, MdS, and H Taheri. 2020. In situ process monitoring for additive manufacturing through acoustic techniques. *Journal of Materials Engineering and Performance* 29(10):6249–62. https://doi.org/10.1007/s11665-020-05125-w.

Kathryn, M, G Moroni, T Vaneker, G Fadel, RI Campbell, I Gibson, A Bernard, et al. 2016. CIRP annals – Manufacturing technology design for additive manufacturing: Trends, opportunities, considerations, and constraints. *CIRP Annals – Manufacturing Technology* 65(2):737–60. https://doi.org/10.1016/j.cirp.2016.05.004.

Khaing, MW, JYH Fuh, and L Lu. 2001. Direct metal laser sintering for rapid tooling: Processing and characterisation of EOS parts. *Journal of Materials Processing Technology* 113(1–3):269–72.

Khanzadeh, M, S Chowdhury, L Bian, and MA Tschopp. 2017. A methodology for predicting porosity from thermal imaging of melt pools in additive manufacturing thin wall sections. *International Manufacturing Science and Engineering Conference*, 50732:V002T01A044.

Khanzadeh, M, S Chowdhury, MA Tschopp, HR Doude, M Marufuzzaman, and L Bian. 2019. In-situ monitoring of melt pool images for porosity prediction in directed energy deposition processes. *IISE Transactions* 51(5):437–55.

Koester, LW, H Taheri, TA Bigelow, LJ Bond, and EJ Faierson. 2018. In-situ acoustic signature monitoring in additive manufacturing processes. *AIP Conference Proceedings* 1949:20006.

Kruth, J-P, X Wang, T Laoui, and L Froyen. 2003. Lasers and materials in selective laser sintering. *Assembly Automation*.

Kwon, O, HG Kim, MJ Ham, W Kim, GH Kim, JH Cho, NI Kim, and K Kim. 2020. A deep neural network for classification of melt-pool images in metal additive manufacturing. *Journal of Intelligent Manufacturing* 31(2):375–86. https://doi.org/10.1007/s10845-018-1451-6.

Lasi, H, P Fettke, H-G Kemper, T Feld, and M Hoffmann. 2014. Industry 4.0. *Business & Information Systems Engineering* 6(4):239–42.

Le, TN, MH Lee, ZH Lin, HC Tran, and YL Lo. 2021. Vision-based in-situ monitoring system for melt-pool detection in laser powder bed fusion process. *Journal of Manufacturing Processes* 68(PA):1735–45. https://doi.org/10.1016/j.jmapro.2021.07.007.

Leary, M, M Mazur, J Elambasseril, M McMillan, T Chirent, Y Sun, M Qian, M Easton, and M Brandt. 2016. Selective laser melting (SLM) of AlSi12Mg lattice structures. *Materials & Design* 98:344–57.

Lee, C, G Seo, D Kim, M Kim, and JH Shin. 2021. Development of defect detection Ai model for wire + arc additive manufacturing using high dynamic range images. *Applied Sciences (Switzerland)* 11(16). https://doi.org/10.3390/app11167541.

Li, Z, X Liu, S Wen, P He, K Zhong, Q Wei, Y Shi, and S Liu. 2018. In situ 3D monitoring of geometric signatures in the powder-bed-fusion additive manufacturing process via vision sensing methods. *Sensors (Switzerland)*. https://doi.org/10.3390/s18041180.

Li, Y, J Polden, Z Pan, J Cui, C Xia, F He, H Mu, H Li, and L Wang. 2021. A defect detection system for wire arc additive manufacturing using incremental learning. *Journal of Industrial Information Integration*:100291.

Lu, QY, and CH Wong. 2018. Additive manufacturing process monitoring and control by non-destructive testing techniques: Challenges and in-process monitoring. *Virtual and Physical Prototyping*. https://doi.org/10.1080/17452759.2017.1351201.

Luo, M, and YC Shin. 2015. Vision-based weld pool boundary extraction and width measurement during keyhole fiber laser welding. *Optics and Lasers in Engineering* 64:59–70.

Maier, RRJ, WN MacPherson, JS Barton, M Carne, M Swan, JN Sharma, SK Futter, DA Knox, BJS Jones, and S McCulloch. 2011. Fibre optic strain and configuration sensing in engineering components produced by additive layer rapid manufacturing. *SENSORS, 2011 IEEE*:1353–57.

McCann, R, MA Obeidi, C Hughes, É McCarthy, DS Egan, RK Vijayaraghavan, AM Joshi, et al. 2021. In-situ sensing, process monitoring and machine control in laser powder bed fusion: A review. *Additive Manufacturing* 45(April). https://doi.org/10.1016/j.addma.2021.102058.

Mireles, J, S Ridwan, PA Morton, A Hinojos, and RB Wicker. 2015. Analysis and correction of defects within parts fabricated using powder bed fusion technology. *Surface Topography: Metrology and Properties* 3(3). https://doi.org/10.1088/2051-672X/3/3/034002.

Pan, Z, D Ding, B Wu, D Cuiuri, H Li, and J Norrish. 2018. Arc welding processes for additive manufacturing: A review. *Transactions on Intelligent Welding Manufacturing* (2):3–24. https://doi.org/10.1007/978-981-10-5355-9_1.

Patel, DB. 2016. Additive manufacturing – Process, applications and challenges 2(5):883–89. www.ijariie.com.

Plessis, AD, I Yadroitsev, I Yadroitsava, and SG Le Roux. 2018. X-ray microcomputed tomography in additive manufacturing: A review of the current technology and applications. *3D Printing and Additive Manufacturing* 5(3):227–47. https://doi.org/10.1089/3dp.2018.0060.

Qidwai, U, and C-H Chen. 2009. *Digital Image Processing: An Algorithmic Approach with MATLAB*. Chapman and Hall/CRC.

Ramalho, A, TG Santos, B Bevans, Z Smoqi, P Rao, and JP Oliveira. 2021. Effect of contaminations on the acoustic emissions during wire and arc additive manufacturing of 316L stainless steel. *Additive Manufacturing* 51(December 2021):102585. https://doi.org/10.1016/j.addma.2021.102585.

Rodriguez, E, J Mireles, CA Terrazas, D Espalin, MA Perez, and RB Wicker. 2015. Approximation of absolute surface temperature measurements of powder bed fusion additive manufacturing technology using in situ infrared thermography. *Additive Manufacturing* 5(2015):31–9. https://doi.org/10.1016/j.addma.2014.12.001.

Rosli, NA, MR Alkahari, MFB Abdollah, S Maidin, FR Ramli, and SG Herawan. 2021. Review on effect of heat input for wire arc additive manufacturing process. *Journal of Materials Research and Technology* 11:2127–45. https://doi.org/10.1016/j.jmrt.2021.02.002.

Salmi, M. 2021. Additive manufacturing processes in medical applications. *Materials* 14(1):191.

Shevchik, SA, C Kenel, C Leinenbach, and K Wasmer. 2018. Acoustic emission for in situ quality monitoring in additive manufacturing using spectral convolutional neural networks. *Additive Manufacturing* 21:598–604. https://doi.org/10.1016/j.addma.2017.11.012.

Shevchik, SA, G Masinelli, C Kenel, C Leinenbach, and K Wasmer. 2019. Deep learning for in situ and real-time quality monitoring in additive manufacturing using acoustic emission. *IEEE Transactions on Industrial Informatics* 15(9):5194–203. https://doi.org/10.1109/TII.2019.2910524.

Taheri, H, LW Koester, TA Bigelow, EJ Faierson, and LJ Bond. 2019. In situ additive manufacturing process monitoring with an acoustic technique: Clustering performance evaluation using K-means algorithm. *Journal of Manufacturing Science and Engineering* 141(4).

Tapia, G, and A Elwany. 2014. A review on process monitoring and control in metal-based additive manufacturing. *Journal of Manufacturing Science and Engineering, Transactions of the ASME* 136(6):1–10. https://doi.org/10.1115/1.4028540.

Toeppel, T, P Schumann, M-C Ebert, T Bokkes, K Funke, M Werner, F Zeulner, F Bechmann, and F Herzog. 2016. 3D analysis in laser beam melting based on real-time process monitoring. In *Materials Science & Technology Conference*:123–32.

Venturini, G, F Baffa, and G Campatelli. 2022. Wire arc additive manufacturing monitoring system with optical cameras. *Lecture Notes in Mechanical Engineering*:151–70. https://doi.org/10.1007/978-3-030-82627-7_9.

Vimal, KEK, M Naveen Srinivas, and S Rajak. 2021. Wire arc additive manufacturing of aluminium alloys: a review. *Materials Today: Proceedings* 41:1139–1145.

Wu, H, Z Yu, and Y Wang. 2016. A new approach for online monitoring of additive manufacturing based on acoustic emission. In *International Manufacturing Science and Engineering Conference* 49910:V003T08A013.

Xia, C, Z Pan, J Polden, H Li, Y Xu, S Chen, and Y Zhang. 2020a. A review on wire arc additive manufacturing: Monitoring, control and a framework of automated system. *Journal of Manufacturing Systems* 57(August):31–45. https://doi.org/10.1016/j.jmsy.2020.08.008.

Xia, C, Z Pan, S Zhang, J Polden, L Wang, H Li, Y Xu, and S Chen. 2020b. Model predictive control of layer width in wire arc additive manufacturing. *Journal of Manufacturing Processes* 58(July):179–86. https://doi.org/10.1016/j.jmapro.2020.07.060.

Xiong, J, and G Zhang. 2013. Online measurement of bead geometry in GMAW-based additive manufacturing using passive vision. *Measurement Science and Technology* 24(11). https://doi.org/10.1088/0957-0233/24/11/115103.

Xu, F, V Dhokia, P Colegrove, A Mcandrew, A Henstridge, ST Newman, F Xu, V Dhokia, P Colegrove, and A Mcandrew. 2018. Realisation of a multi-sensor framework for process monitoring of the wire arc additive manufacturing in producing Ti-6Al-4V parts. *International Journal of Computer Integrated Manufacturing* 00(00):1–14. https://doi.org/10.1080/09511 92X.2018.1466395.

Xu, P, G Xu, X Tang, and S Yao. 2008. A visual seam tracking system for robotic arc welding. *International Journal of Advanced Manufacturing Technology* 37(1–2):70–75. https://doi.org/10.1007/s00170-007-0939-6.

Yang, D, G Wang, and G Zhang. 2017. A comparative study of GMAW- and DE-GMAW-based additive manufacturing techniques: Thermal behavior of the deposition process for thin-walled parts. *International Journal of Advanced Manufacturing Technology* 91(5–8):2175–84. https://doi.org/10.1007/s00170-016-9898-0.

Zhang, Y, GS Hong, D Ye, K Zhu, and JYH Fuh. 2018a. Extraction and evaluation of melt pool, plume and spatter information for powder-bed fusion AM process monitoring. *Materials and Design* 156:458–69. https://doi.org/10.1016/j.matdes.2018.07.002.

Zhang, B, S Liu, and YC Shin. 2019. In-process monitoring of porosity during laser additive manufacturing process. *Additive Manufacturing* 28:497–505. https://doi.org/10.1016/j.addma.2019.05.030.

Zhang, Z, Z Liu, and D Wu. 2021. Prediction of melt pool temperature in directed energy deposition using machine learning. *Additive Manufacturing* 37:101692. https://doi.org/10.1016/j.addma.2020.101692.

Zhang, Y, L Wu, X Guo, S Kane, Y Deng, YG Jung, JH Lee, and J Zhang. 2018b. Additive manufacturing of metallic materials: A review. *Journal of Materials Engineering and Performance* 27(1). https://doi.org/10.1007/s11665-017-2747-y.

Zhao, J, R Xia, W Liu, and H Wang. 2009. A computing method for accurate slice contours based on an STL model. *Virtual and Physical Prototyping* 4(1):29–37.

Chapter 5

Additive manufacturing

Materials, technologies, and applications

Raj Agarwal, Shrutika Sharma,
Vishal Gupta, and Jaskaran Singh
Thapar Institute of Engineering and Technology

Kanwaljit Singh Khas
Lovely Professional University

CONTENTS

5.1	Introduction	77
5.2	Materials and technologies used in AM	79
	5.2.1 Polymers	79
	5.2.1.1 Fused deposition modelling (FDM) technique	79
	5.2.1.2 Polyjet 3D printing technique	81
	5.2.1.3 Stereolithography (SLA) technique	81
	5.2.1.4 Selective laser sintering (SLS) technique	83
	5.2.2 Ceramics	84
	5.2.2.1 Multi-jet printing	87
	5.2.2.2 Extrusion printing	87
	5.2.2.3 Jet printing	88
5.3	Recent advancements and possibilities of additive manufacturing technologies for various applications	89
	5.3.1 Application of 3D printing in consumer products	89
	5.3.2 Application of 3D printing in aerospace industry	89
	5.3.3 Application of 3D printing in food industry	90
	5.3.4 Application of 3D printing in biomedical domain	90
	5.3.5 Application of 3D printing in automobile industry	92
5.4	Limitations and future scope	93
5.5	Conclusion	93
References		94

5.1 INTRODUCTION

Conventional processes of manufacturing result in huge wastage of material, consumption of time, and loss of precision. This limitation of conventional processes can be overcome by additive manufacturing (AM), which involves layer-by-layer deposition of material resulting in faster

DOI: 10.1201/9781003258391-5

part building and production of customized parts with minimized waste material [1]. The three-dimensional gels of the part are made in designing software, which is then tessellated to form a standard tessellation language (STL) file and printing of the part takes place with the deposition of each layer (Figure 5.1). This technique holds the ability to fabricate parts with complex and intricate geometries with minimum requirements for postprocessing treatments. A variety of materials ranging from metals to polymers and ceramics can be used to produce unique parts at low numbers in an economical manner [2,3]. Another factor that drives the trend of AM is its promising ecological and environmental behaviour. A wide variety of AM processes are currently available that differ from each other in terms of layer deposition principle and variety of materials that can be used. Figure 5.2 illustrates different materials and technologies used in AM technology for the fabrication of parts. Polymers such as polycaprolactone (PCL), polylactic acid (PLA), and polyglycolic acid (PGA) are mainly manufactured by polyjet 3D printing, fused deposition modelling, stereolithography, and

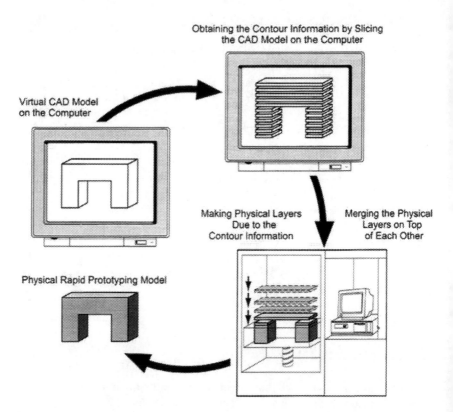

Figure 5.1 Steps of processes involved in additive manufacturing. (Reproduced with permission from Elsevier [4].)

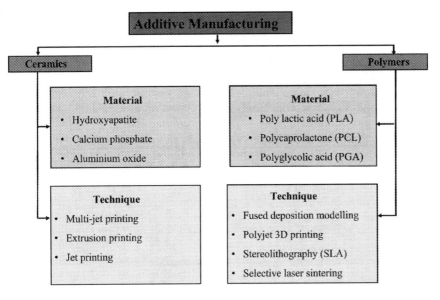

Figure 5.2 Additive manufacturing used materials and techniques.

selective laser sintering. AM processes like multi-jet printing, extrusion printing, and jet printing are mainly used for manufacturing ceramic parts. Each of these processes has been explained in detail in subsequent sections. This chapter gives a glance at the most preferable polymers and ceramics that are used as base materials for AM processes. Apart from this, the variety of AM processes that are used for part fabrication using polymers and ceramics have also been explained. Briefly, this chapter highlights the potential of AM in the fabrication of consumer products, aerospace, food, and biomedical and automobile parts.

5.2 MATERIALS AND TECHNOLOGIES USED IN AM

5.2.1 Polymers

5.2.1.1 Fused deposition modelling (FDM) technique

FDM or fused filament fabrication (FFF) technology is a technique in which a thermoplastic polymer is extruded through a heated extrusion head that heats the polymer to a semi-molten state and immediately bonds to the lay-down pattern on a build stage [5,6]. The thermoplastic polymer is deposited on the build plate in the semi-liquid state with the help of two nozzles [7]. The layers are fused to build the part layer by layer, as depicted in Figure 5.3 [8,9]. The material must have a low melting temperature, heat transfer characteristics, and rheology for FDM material selection criteria [10–12]. This

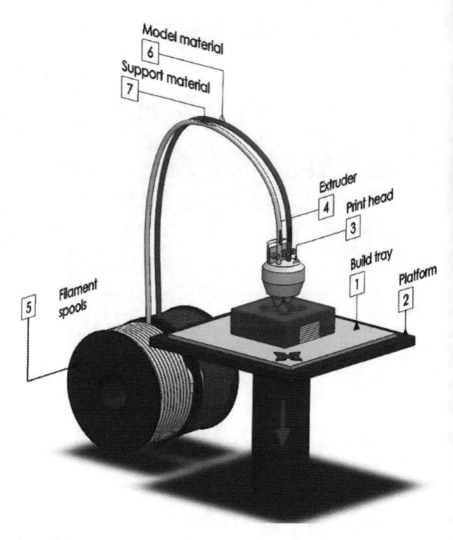

Figure 5.3 Principle of fused deposition modelling machine. (Reproduced with permission from Elsevier [16].)

technology can produce high-porosity models due to the laydown pattern [13]. This technology is the most affordable and budget-friendly AM technology [14]. The laydown pattern providing high porosity along with good mechanical properties distinguishes FDM from other processes. However, this process is limited to thermoplastic materials possessing viscosity that is low for the extrusion process and high enough for structure building. FDM is not limited to the geometrical complexity of structures using industrial materials with optimal thermal and rheological properties [15]. The key disadvantage of FDM involves its inability to incorporate living cells during the process of extrusion as a high processing temperature is required for the extrusion process which might result in the killing of living cells.

5.2.1.2 Polyjet 3D printing technique

Polyjet technology works similarly to traditional inkjet printing. A carriage with four or more inkjet heads deposits tiny droplets of photopolymer in a liquid state on the build tray and ultraviolet (UV) lamp that solidifies the material [17]. Multiple jetting heads allow different materials (in multi-colour format) to be ejected at the same time enabling this technology to produce the object with various levels of flexibility [18,19]. With a minimum wall thickness of 1 mm, microscopic layer resolution of 16 microns (0.016 mm) and accuracy up to 0.1–0.3 mm can be obtained [20]. The polyjet 3D printer offers polished, smooth surface, exceptionally detailed and precise prototypes using the broadest range of materials with accuracy, precision, and excellent detailing [21,22]. This technology can achieve complex geometries, delicate features, and intricate details. The main part of this technology is the usage of the widest variety of colours and materials in an individual model. An essential application of polyjet 3D printers is in the healthcare sector that enables surgeons to produce a replica of organs that need to be replaced. Prosthetic limbs, joint replacement, and customized implant fabrication are the potential of this technology. The process can produce structures with complicated geometries and thin walls such as implant models, realistic teeth, and maxillofacial models with excellent accuracy, detailing, and precision.

The polyjet 3D printer offers polished, smooth surfaces, and exceptionally detailed prototypes that can convey the aesthetics of the final product. It helps in the manufacturing of fixtures, jigs, and tools with high accuracy. This technology can fabricate highly complex shapes with delicate features and intricate details. The technique holds the primary feature of usage of a variety of materials and colours into a single model with high efficiency.

5.2.1.3 Stereolithography (SLA) technique

SLA was the first AM process to be theorized and was developed in the late 1980s. SLA uses a photopolymer resin with a thin layer that a UV laser can cure [19]. The resin solidifies into a single two-dimensional layer of the computer aided design (CAD) model. After every layer is cured, the platform moves down in the bottom-up approach with the attached cured structure and other layers of uncured resin spread over the top for solidification [21]. After building the model, the non-polymerized liquid resin can be collected and reused by draining. The working principle of the SLA machine is illustrated in Figure 5.4. For large parts, curing an entire layer of photopolymers at the same time requires the masked lamp technique. Draining helps in

82 Additive Manufacturing

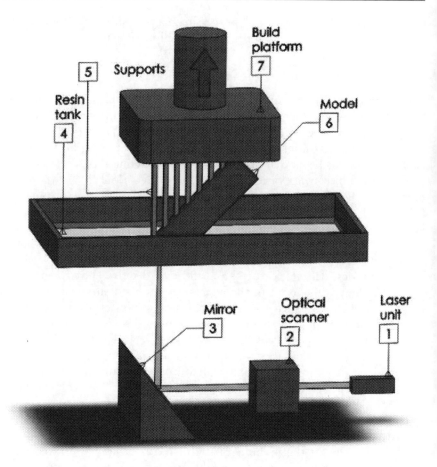

Figure 5.4 Principle of stereolithography. (Reproduced with permission from Elsevier [16].)

the removal of liquid non-polymerized resin after building the structure. Kinetics of curing reactions is a critical area that affects layer thickness polymerized and curing time. This can be controlled with the help of scanning speed, light source power, amount of monomer, and chemistry present within the monomers. Apart from this, the depth of polymerization can be controlled by the addition of UV absorbers.

The advantages of this fast fabrication process of SLA technology are creating complex shapes with auxetic internal structure, the comfort of abstraction of unpolymerized resin, functional prototyping, patterns, and high-resolution part (~0.2 μm) with accuracy [20]. The drawback of SLA technology is that photo-polymerized resin with poor mechanical properties essentially is provided for tissue engineering. Comprehensive support structure removal may be difficult. Biomedical applications of SLA embrace the indirect fabrication of medical devices, pre-surgical planning, and fabrication of anatomical models using the SLA patterns for casts. Titanium dental implants have been made up of electrical discharge machining (EDM) of titanium ingot based on an SLA model.

5.2.1.4 Selective laser sintering (SLS) technique

In the process, a laser source with high power is used for the fusion of small metallic, ceramic, thermoplastic particles, or glass powders [19]. The surface of polymer powdered particles is scanned by laser in a two-dimensional pattern for heating of powder above the glass transition temperature. After the deposition of every layer, the bed is lowered and a fresh layer of powdered material is deposited over the top surface [19]. The working principle of the SLA machine is illustrated in Figure 5.5. The un-sintered, unbound, loose powder works as a support material that is removed after the part is completed. In practice, due to the single melting point of all powders, the melting is relatively easy to achieve. Also, with the detailed resolution, no extra support is required as the powder works as the support material [23,24]. Materials for the SLS process are limited to thermoplastic polymers [25]. The incredible resolution of features is determined by the particle size of powder, the diameter of a focused laser beam, and heat transfer in

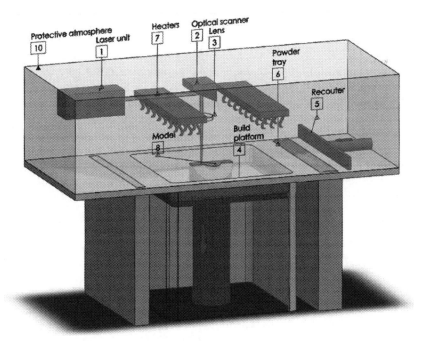

Figure 5.5 Principle of selective laser sintering. (Reproduced with permission from Elsevier [16].)

84 Additive Manufacturing

the powder bed. Moreover, fast sintering and poor spreading of powder may cause edge inaccuracies. The porosity in models fabricated with SLS technology is limited as the particle size of the powder may vary the size of pores in the model. The drawback of too small particles of powder material may result in poor spreading and can also cause powder clumping [25]. A comparison of polymer-based AM technologies is presented in Table 5.1.

The surface of powdered particles is scanned by laser to sinter by heating them above the glass transition temperature. Sintering results in molecular diffusion along the particle's outermost surface lead to neck formation between neighbouring particles. After deposition of one layer, the piston with the fabricated part is lowered and fresh powdered material is rolled across the surface [19]. This leads to the formation of a subsequent layer strongly bonded to the previous layer. The loose powder gets removed once the part fabrication is completed and full density is achieved after heat treatment. Since sintering does not cause complete melting of powdered particles, the porosity present between original particles can be maintained.

In the case of pure metals, melting of the entire powder takes place at a single temperature, whereas it is difficult to accomplish melting in the case of alloys due to variation in surface tension, liquid flow behaviour, and interactions between material and laser. Materials for the SLS process are limited to thermoplastic polymers [25]. The particle size of powder, heat transfer on powdered bed, and diameter of focused laser beam play an important role in determining the resolution of features of this process. Quick sintering and poor spreading of powder result in edge inaccuracies, so the particle size is limited to 10 µm. The most used materials are PCL and a combination of hydroxyapatite (HA) and polyether ether ketone (PEEK). Thin solid disks are made with biomaterials on the ~400–500 µm scale.

Fabrication of scaffolds with anatomically shaped external structures and the porous interior becomes easy with the SLS Technique. Medical grade PEEK has been awarded FDA clearance for the fabrication of craniofacial implants using the SLS process. However, this process cannot be used for materials with a smaller pore size as created pores highly depend on powder particle size. Poor spreading from the clumping of powder is another reason for limiting powder particles to a large size [25].

5.2.2 Ceramics

Ceramic material may be termed as any inorganic crystalline material, consisting of a metallic element along with non-metallic elements such as the bonding of aluminium with non-metallic elements through ionic or covalent bonding. Calcium phosphate (CaP) ceramic is highly used in biomedical applications due to its composition being close to bone properties [26]. Commonly used ceramics include HA [27,28], alumina [28], silica [29], zirconia [30], and calcium sulphate [31]. Tricalcium phosphate (TCP)

Table 5.1 A comparison of polymer-based additive manufacturing technology

Observation	Polyjet 3D printer	Fused deposition modelling(FDM)	Stereolithography(SLA)	Selective laser sintering(SLS)	Reason
Operations					
Type	UV cured extrusion	Material extrusion	Vat polymerization	Powder bed fusion	The operation techniques of various polymer-based additive manufacturing technologies use different binding strategies to fabricate the final product from the 3D CAD model.
Process time	↑↑↑	↑↑	↑↑	↑↑	
Pre-process	↑↑	↑↑↑	↑↑	↑	
Post-process	↑↑↑	↑↑↑	↑↑	↑↑↑	
Office environment	↑↑↑	↑↑↑	↑↑↑	↑↑↑	
Ease of use	↑↑↑	↑↑↑↑	↑↑↑	↑↑↑↑	
Parameters					
Layer thickness (mm)	0.1–0.3	0.05–0.127	0.05–0.015	0.05–0.01	Every technique has its concept of fabrication, which is limited to various input parameters and binding volumes.
Minimum wall thickness (mm)	1	1	5	0.8	
Build volume (mm)	294×192×148.6	1000×1000×1000	1500×650×550	550×550×750	
Support (complex design)	Required	Required	Required	Not required	

(Continued)

Table 5.1 (Continued) A comparison of polymer-based additive manufacturing technology

Observation	Polyjet 3D printer	Fused deposition modelling(FDM)	Stereolithography(SLA)	Selective laser sintering(SLS)	Reason
Characteristics					
Surface finish	↑↑↑	↑	↑↑↑↑	↑↑↑↑	The characteristics of various techniques have different outputs based on their usage and costing.
Feature detail	↑↑↑	↑	↑↑↑↑	↑↑↑	
Accuracy	↑↑	↑↑↑	↑↑↑↑	↑↑↑	
Size	↑	↑↑↑	↑↑↑↑	↑↑	
Cost	↑↑	↑	↑↑	↑↑↑	
Materials					
Rigid	↑↑↑	↑↑↑	↑↑↑	↑↑↑	Different techniques consume different materials (may use different colours as well) based on the application of usage.
Flexible	↑↑↑	↑	↑	↑	
Durable	↑↑	↑↑↑	↑↑↑↑	↑↑↑	
Transparent	↑↑↑	↑	↑	↑↑	
High performance	↑	↑↑↑	↑↑↑↑	↑↑↑↑	
Bio-compatible	↑↑↑	↑↑↑	↑↑↑	↑↑↑	

is preferred over HA due to its better biodegradability. Bioglass ceramics possess better bioactivity [30] and 45S5 bioglass is also used for the fabrication of scaffolds for bone tissue engineering; however, it does have the ability to withstand loads [31,32]. Coating of the implant with a bioactive ceramic such as HA and bioactive glass helps in improving osseointegration. The coating can also be done with plasma spraying, laser ablation, electrophoretic deposition, and ionic sputtering; however, they are not cost-effective [27,28,31,33].

5.2.2.1 Multi-jet printing

Multi-jet printing is a process that brings piezo print head technology into usage for deposition of casting wax material or photocurable resin in a layer-by-layer manner. The layer thickness is typically 80 μm, the minimum wall thickness is 1 mm with standard accuracy of ±0.3%, and the minimum feature size is 0.5 mm. This technique involves the usage of an inkjet array for selective application of detailing and fusing agents across a bed laid with nylon powder that is fused by heating elements into a single solid layer. After deposition of each layer, the powder is distributed on top of the bed and this process repeats until the fabrication of the part is completed. On finishing the part fabrication, the powder bed with encapsulated parts is transported to a processing station where an integrated vacuum is used to remove the loose powder. Bead blasting of parts takes place for removal of residual powder before the part is transported to the finishing department where parts are dyed black for improvement in appearance.

The advantages of this process include accelerated build speed and build time so it is more economical, fine minimum feature size, low porosity, best surface resolution details with improved surface roughness, fully dense, consistency in isotropic mechanical properties in all three build directions as compared to other additive processes. This technology is ideal for the fabrication of functional prototypes with a short lead time, reduced porosity, and excellent surface quality.

5.2.2.2 Extrusion printing

There are mainly two classes of extrusion printing in AM. One is FDM which is the most popular process that mainly uses plastics and polymeric matrix composition polymers. Another technology is paste deposition modelling (PDM) which is an extrusion method using a filament of highly viscous pastes and ceramics [34]. The process is quite similar to that of FDM of polymers, but the used paste is extruded and deposited at room temperature and solidified by evaporation of water [35]. The 3D model is built from the bottom layer to the top layer with deposition of one layer at a time. A thin layer of ceramic powder is spread on the platform using the printer's rolling mechanism organic binder that is placed at specific locations by the print head. The nozzles move from one place to another for building up objects in a layer-by-layer manner until the fabrication of a part is completed [35].

PDM technology can replace the manufacturing of ceramics as compared to the traditional manufacturing methods. The advantages of PDM are that it is a cheap and fast technology that eliminates the construction periods of the models, moulds, and matrices; provides extension of morphological possibilities; and ensures low production cost of individual pieces or small series [34]. The semi-liquid paste helps in the retention of shape, resulting in improved viscoelastic and loading properties [35].

5.2.2.3 Jet printing

Jet printing of ceramics can be achieved with two methods, i.e., binder jetting and material jetting. It involves the distribution of a thin powdered layer on a platform with the help of a levelling roller. It consists of an ink-jet nozzle that moves in x and y axes for the distribution and adherence of powder. After deposition of one powdered layer, the bed moves down by some distance on the z-axis followed by the distribution of another powdered layer and the process for binder injection repeats. After part building, the prepared part is taken for post-processing, as depicted in Figure 5.6.

Better surface quality with the help of micro-droplets can be achieved with this process and can be used for the fabrication of overhanging structures using sacrificial materials for applications in microfluidics and structural components [25]. However, expensive commercial printers and proprietary resins are key limitations of this process.

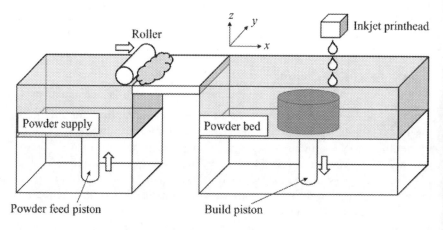

Figure 5.6 Binder jet 3D printing. (Reproduced with permission from Elsevier [36].)

5.3 RECENT ADVANCEMENTS AND POSSIBILITIES OF ADDITIVE MANUFACTURING TECHNOLOGIES FOR VARIOUS APPLICATIONS

The 3D printing–based AM technologies have recently been cultivated and advanced in various domains in different sectors. This section presents recent advancements and possibilities of AM technologies for various industrial applications such as consumer products, aerospace industry, automobile industry, food industry, and biomedical domain. In the medical domain, our focus is on orthopaedics, and in the non-medical domain, the focus is on the food industry, automotive industry, consumer products, and aerospace industry. The usage and real-world application of AM technologies were found to be adaptable and progressive.

5.3.1 Application of 3D printing in consumer products

The use of 3D printing for the fabrication of consumer products helps in the reduction of waste in processes of production and consumption through reconfiguration of additive manufactured parts. Through this system, custom parts that can be 3D printable can be found online by the consumers. This process helps in the elimination of complicated 3D modelling, which is hardly capable of modelling new custom parts. This helps in the elimination of unnecessary repetition of products with similar functionalities and the creative needs of consumers are encouraged by introducing diverse features in products depending on consumers' needs [37]. The main advantage of 3D printing consumer products is the provision of satisfaction according to the interests of consumers. The model has its online availability where the user can edit the model for best fit and personalized usage, mainly depending on structural and functional requisites. It provides a configuration in which consumers and producers stay connected by giving the freedom to consumers for modification and extension of the products.

5.3.2 Application of 3D printing in aerospace industry

In 2015, the first 3D printed part was fabricated for aerospace applications that were made of silver for providing housing to the inlet temperature sensor of the compressor, present inside a jet engine. LEAP (leading-edge aviation propulsion) engines are currently using 19 3D printed fuel nozzles. Titanium leading-edge blades are being planned to be replaced with 3D printed parts as proposed by General Electric Aviation. Air cooling ducts in Super Hornet Jets and parts of Bell 429 Helicopter are being made from the 3D Printing process. Boeing is fabricating plastic interior parts with the help of the 3D Printing process. Ultem and nylon are being used for

90 Additive Manufacturing

the fabrication of prototypes and tooling to produce composite parts. The future of aerospace industries in terms of 3D printing lies in the opportunity of using nanocomposites in which nanoparticles with excellent mechanical and thermal properties can be added to the host material matrix which will be 3D printed into nanocomposite parts. Unmanned air vehicles and experimental aircraft can be fabricated with the help of 3D printing that requires the least regulatory scrutiny. High energy demand and a set of certification rules are some of the challenges that need to be addressed for the complete and successful implementation of 3D printing in aerospace industries [38]. The adoption of 3D printing technology for the aerospace industry is still slow due to the requirement for further development in areas of testing and standards for safety. It is expected that 3D printing technology has the potential for various applications in the aerospace industry.

5.3.3 Application of 3D printing in food industry

In today's world, AM not only works in mechanical workshops but is also used in the kitchen to make good-quality food. Manufacturing complex parts economically at low volume through AM has led to its increased usage in food applications. Complex textures with artistic presentation can be imparted to geometrically complex parts. 2D printable artwork present on cakes is now being used for shaping internal and external features of food [39]. The shape of food can be customized based on the choice of customers by the addition of premium food products. 3D printing has been used to produce sugar sculptures by using the ink-jetting process on powdered sugar layers, as illustrated in Figure 5.7. Food items like pizza, cake, etc. can be made through 3D manufacturing technology. In food processing industries, different shapes and sizes of cakes can be easily baked or produced using AM techniques. For 3D printing of pizza, thickness, dough, cheese, and sauce are controlled in the layering process of 3D Printing, as demonstrated in Figure 5.8. It is expected that 3D printing technology may be used and implemented on an industrial scale for the manufacturing of food.

The main limitation in food processing is the dependence on the formulation of material, mechanical force, the proper blending of materials, and stability of material supply. Piracy of digital recipes is a major challenge that requires policies for protection against intellectual rights. Machine vision systems and food quality evaluation systems need to be in place for the successful implementation of food printing. This can help in the manufacturing of nutrient-dense food as per the individual requirements of the common man.

5.3.4 Application of 3D printing in biomedical domain

A bone fracture may happen due to unintentional accidents, stress fracture, and ageing [41–43]. The fixation of the fractured part of the bone is done

Additive manufacturing 91

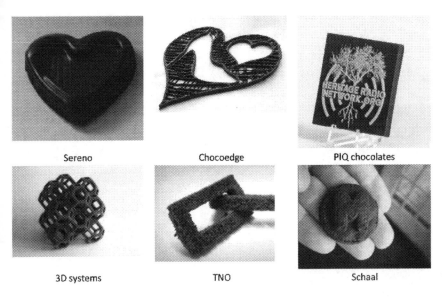

Figure 5.7 Application of 3D printing technology in food industries. (Reproduced with permission from Elsevier [40].)

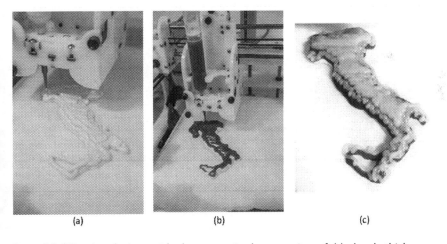

Figure 5.8 3D printed pizza with the customized proportion of (a) dough thickness, (b) sauce, and (c) cheese. (Reproduced with permission from Elsevier [40].)

with complex orthopaedic surgery. AM technologies can be used for various applications during surgery. The 3D printing technology can be utilized for preoperative planning as the surgeon may perform the surgery on the 3D printed parts to save the surgery time and blood loss of the patient. A study of external fixators is presented in Figure 5.9. The FDM-based 3D

92 Additive Manufacturing

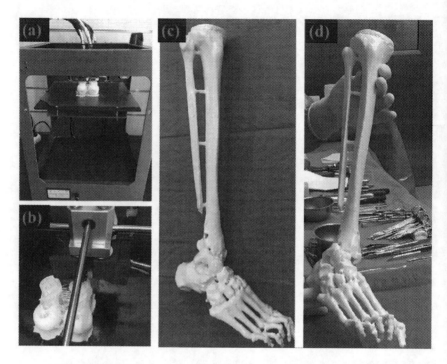

Figure 5.9 (a) Fused deposition modelling 3D printer, (b) fabrication of patient model using polylacticacid polymer, (c) 3D printed model, and (d) sterilized printed model used during surgery. (Reproduced with permission from Elsevier [44].)

printing technology is used for the fabrication of patient-specific models for planning the surgery, as illustrated in Figure 5.9c. It is expected that 3D printing technology has the potential that can be used for preoperative planning, accurate physical rapid prototyping of patient-specific models, and surgical planning. Moreover, various implantable devices can also be fabricated with AM technologies for better fracture fixation of the fractured bone.

5.3.5 Application of 3D printing in automobile industry

The application of AM technologies in the automobile industry may help in fabricating complicated parts with new aesthetic designs and complicated shapes. The quick processing and desired design based on imagination are the key factors of using AM techniques in the automobile industry. Various ranges of materials can be explored to reduce the weight of the vehicle. The AM technologies are time-saving processes for the fabrication of various parts used in vehicles. 3D printings of front knuckle and ball

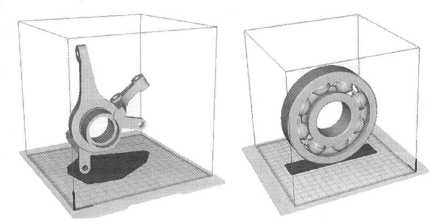

Figure 5.10 3D printing of front knuckle and ball bearing meshing of vehicle. (Reproduced with permission from Elsevier [45].)

bearing meshing for the automobile vehicles are illustrated in Figure 5.10. The weight reduction and complicated geometrical parts fabrication with AM technologies can be a vital solution for use of this technology in the automobile industry.

5.4 LIMITATIONS AND FUTURE SCOPE

High demand and growth in the application of AM technology for various industrial applications have been observed in recent times. The processing defects, mechanical strength, and inconsistent product quality may disrupt the use of AM technology for industrial applications. The capabilities and real potential of rapid prototyping technologies can be enhanced using data-driven intelligent approaches with the adaptation of machine learning and artificial intelligence. The prediction and optimization of independent variables and output responses of additive manufactured parts may enhance the industrial use of AM technology.

5.5 CONCLUSION

AM or rapid prototyping technology provides flexibility and accuracy in part designing and offers fast production along with minimization of by-product wastage. Apart from these advantages, it is cost and environment friendly with the provision of a variety of materials according to the customer requirements. It involves a variety of processes that can be used depending on the state of the material. With the possibility of manufacturing structures with lighter weight, AM is being used in aerospace industries as fabrication of aerospace parts with reduced weight is the common goal of these industries. In automobile industries, AM has been successful in the fabrication of vehicle parts with intricate and complex geometries. This technique has taken the biomedical field to new heights with the fabrication of bones and implants before surgery which has helped in preoperative assessment and better communication between the surgeon and patient through real surgical models. AM has provided an easier way for manufacturing food with a variety of 3D printed food tools for customer-specific food designs. Evidently, with continuous growth and successful results, AM holds the future of manufacturing. However, further research is still required as every common manufacturing material cannot be used for AM techniques. Moreover, higher levels of precision and accuracy need to be achieved to eliminate the requirement of post-processing for AM.

REFERENCES

1. Tagscherer N, Consul P, Kottenstedde IL, et al. (2021) Investigation of nonisothermal fusion bonding for extrusion additive manufacturing of large structural parts. *Polym Compos* 42:5209–5222 https://doi.org/10.1002/pc.26216.
2. Agarwal R, Gupta V, Singh J (2022) Mechanical and biological behaviour of additive manufactured biomimetic biodegradable orthopaedic cortical screws. *Rapid Prototyp J* 4:1–25 https://doi.org/10.1108/rpj-01-2022–0006.
3. Agarwal R, Gupta V, Singh J (2022) Additive manufacturing-based design approaches and challenges for orthopaedic bone screws: A state-of-the-art review. *J Brazilian Soc Mech Sci Eng* 44:1–25 https://doi.org/10.1007/s40430-021-03331–8.
4. Prakash KS, Nancharaih T, Rao VVS (2018) Additive manufacturing techniques in manufacturing – An overview. *Mater Today Proc* 5:3873–3882 https://doi.org/10.1016/j.matpr.2017.11.642.
5. Dhandapani R, Krishnan PD, Zennifer A, et al. (2020) Additive manufacturing of biodegradable porous orthopaedic screw. *Bioact Mater* 5:458–467 https://doi.org/10.1016/j.bioactmat.2020.03.009.
6. Daminabo SC, Goel S, Grammatikos SA, et al. (2020) Fused deposition modeling-based additive manufacturing (3D printing): Techniques for polymer material systems. *Mater Today Chem* 16:100248 https://doi.org/10.1016/j.mtchem.2020.100248.
7. Francis V, Jain PK (2018) A filament modification approach for in situ ABS/OMMT nanocomposite development in extrusion-based 3D printing. *J Brazilian Soc Mech Sci Eng* 40:1–13 https://doi.org/10.1007/s40430-018-1282–6.
8. Chia HN, Wu BM (2015) Recent advances in 3D printing of biomaterials. *J Biol Eng* 9:1–14 https://doi.org/10.1186/s13036-015-0001–4.
9. Mago J, Kumar R, Agrawal R, et al. (2020) Modeling of linear shrinkage in PLA parts fabricated by 3D printing using TOPSIS method. In: Shunmugam MS and Kanthababu M (eds) *Advances in Additive Manufacturing and Joining.* Springer, Singapore, pp 267–276.

10. Guessasma S, Zhang W, Zhu J, et al. (2015) Challenges of additive manufacturing technologies from an optimisation perspective. *Int J Simul Multidiscip Des Optim* 6:A9 https://doi.org/10.1051/smdo/2016001.
11. Guo N, Leu MC (2013) Additive manufacturing: Technology, applications and research needs. *Front Mech Eng* 8:215–243 https://doi.org/10.1007/s11465-013-0248-8.
12. Agarwal R, Mehtani HK, Singh J, Gupta V (2022) Post-yielding fracture mechanics of 3D printed polymer-based orthopedic cortical screws. *Polym Compos* 43:1–9 https://doi.org/10.1002/pc.26620.
13. Korkut V, Yavuz H (2020) Enhancing the tensile properties with minimal mass variation by revealing the effects of parameters in fused filament fabrication process. *J Brazilian Soc Mech Sci Eng* 42:1–18 https://doi.org/10.1007/s40430-020-02610-0.
14. Agarwal R, Malhotra S, Gupta V, Jain V (2022) The application of three-dimensional printing on foot fractures and deformities: A mini-review. *Ann 3D Print Med* 5:100046 https://doi.org/10.1016/j.stlm.2022.100046.
15. Jain R, Nauriyal S, Gupta V, Khas KS (2021) Effects of process parameters on surface roughness, dimensional accuracy and printing time in 3D printing. In: *Advances in Production and Industrial Engineering.* Springer, pp 187–197.
16. Szymczyk-Ziółkowska P, Łabowska MB, Detyna J, et al. (2020) A review of fabrication polymer scaffolds for biomedical applications using additive manufacturing techniques. *Biocybern Biomed Eng* 40:624–638 https://doi.org/10.1016/j.bbe.2020.01.015.
17. Kumar R, Kumar M, Chohan JS (2021) The role of additive manufacturing for biomedical applications: A critical review. *J Manuf Process* 64:828–850 https://doi.org/10.1016/j.jmapro.2021.02.022.
18. Justin Stiltner L, Elliott AM, Williams CB (2011) A method for creating actuated joints via fiber embedding in a polyjet 3D printing process. *22nd Annu Int solid Free Fabr Symp*, pp 583–592.
19. Meng C, Ho B, Ng SH, Yoon Y (2015) A review on 3D printed bioimplants. *Int J Precis Eng Manuf* 16:1035–1046 https://doi.org/10.1007/s12541-015-0134-x.
20. Dommati H, Ray SS, Wang JC, Chen SS (2019) A comprehensive review of recent developments in 3D printing technique for ceramic membrane fabrication for water purification. *RSC Adv* 9:16869–16883 https://doi.org/10.1039/c9ra00872a.
21. Ganguli A, Pagan-Diaz GJ, Grant L, et al. (2018) 3D printing for preoperative planning and surgical training: A review. *Biomed Microdevices* 20 https://doi.org/10.1007/s10544-018-0301-9.
22. Agarwal R (2021) The personal protective equipment fabricated via 3D printing technology during COVID-19. *Ann 3D Print Med* 5:100042 https://doi.org/10.1016/j.stlm.2021.100042.
23. Rendeki S, Nagy B, Bene M, et al. (2020) An overview on personal protective equipment (PPE) fabricated with additive manufacturing technologies in the era of COVID-19 pandemic. *Polymers (Basel)* 12:1–18 https://doi.org/10.3390/polym12112703.
24. Singh S, Prakash C, Ramakrishna S (2019) 3D printing of polyether-ether-ketone for biomedical applications. *Eur Polym J* 114:234–248 https://doi.org/10.1016/j.eurpolymj.2019.02.035.

25. Karakurt I, Lin L (2020) 3D printing technologies: Techniques, materials, and post-processing. *Curr Opin Chem Eng* 28:134–143 https://doi.org/10.1016/j.coche.2020.04.001.
26. Butscher A, Bohner M, Doebelin N, et al. (2013) New depowdering-friendly designs for three-dimensional printing of calcium phosphate bone substitutes. *Acta Biomater* 9:9149–9158 https://doi.org/10.1016/j.actbio.2013.07.019.
27. Will J, Melcher R, Treul C, et al. (2008) Porous ceramic bone scaffolds for vascularized bone tissue regeneration. *J Mater Sci Mater Med* 19:2781–2790 https://doi.org/10.1007/s10856-007-3346-5.
28. Li J, Fartash B, Hermansson L (1995) Hydroxyapatite-alumina composites and bone-bonding. *Biomaterials* 16:417–422 https://doi.org/10.1016/0142-9612(95)98860-G.
29. Zocca A, Elsayed H, Bernardo E, et al. (2015) 3D-printed silicate porous bioceramics using a non-sacrificial preceramic polymer binder. *Biofabrication* 7:025008 https://doi.org/10.1088/1758-5090/7/2/025008.
30. Ke D, Bose S (2018) Effects of pore distribution and chemistry on physical, mechanical, and biological properties of tricalcium phosphate scaffolds by binder-jet 3D printing. *Addit Manuf* 22:111–117 https://doi.org/10.1016/j.addma.2018.04.020.
31. Moore WR, Graves SE, Bain GI (2001) Synthetic bone graft substitutes. *ANZ J Surg* 71:354–361 https://doi.org/10.1046/j.1440-1622.2001.02128.x.
32. Chen Z, Li Z, Li J, et al. (2019) 3D printing of ceramics: A review. *J Eur Ceram Soc* 39:661–687 https://doi.org/10.1016/j.jeurceramsoc.2018.11.013.
33. Najeeb S, Khurshid Z, Matinlinna JP, et al. (2015) Nanomodified peek dental implants: Bioactive composites and surface modification – A review. *Int J Dent* 2015 https://doi.org/10.1155/2015/381759.
34. Ruscitti A, Tapia C, Rendtorff NM (2020) A review on additive manufacturing of ceramic materials based on extrusion processes of clay pastes. *Ceramica* 66:354–366 https://doi.org/10.1590/0366-69132020663802918.
35. Nair NR, Sekhar VC, Nampoothiri KM, Pandey A (2017) Biodegradation of biopolymers. In: Pandey A, Negi S, Soccol CR (eds) *Current Developments in Biotechnology and Bioengineering*. Elsevier, pp 739–755.
36. Zhang Y, Jarosinski W, Jung YG, Zhang J (2018) *Additive Manufacturing Processes and Equipment*. Elsevier Inc.
37. Yoo B, Ko H, Chun S (2016) Prosumption perspectives on additive manufacturing: Reconfiguration of consumer products with 3D printing. *Rapid Prototyp J* 22:691–705 https://doi.org/10.1108/RPJ-01-2015-0004.
38. Joshi SC, Sheikh AA (2015) 3D printing in aerospace and its long-term sustainability. *Virtual Phys Prototyp* 10:175–185 https://doi.org/10.1080/17452759.2015.1111519.
39. Nachal N, Moses JA, Karthik P, Anandharamakrishnan C (2019) Applications of 3D printing in food processing. *Food Eng Rev* 11:123–141 https://doi.org/10.1007/s12393-019-09199-8.
40. Lipton JI, Cutler M, Nigl F, et al. (2015) Additive manufacturing for the food industry. *Trends Food Sci Technol* 43:114–123 https://doi.org/10.1016/j.tifs.2015.02.004.

41. Agarwal R, Jain V, Gupta V, et al. (2020) Effect of surface topography on pull-out strength of cortical screw after ultrasonic bone drilling: An in vitro study. *J Brazilian Soc Mech Sci Eng* 42:1–13 https://doi.org/10.1007/s40430-020-02449-5.

42. Agarwal R, Gupta V, Jain V (2021) A novel technique of harvesting cortical bone grafts during orthopaedic surgeries. *J Brazilian Soc Mech Sci Eng* 8:1–14 https://doi.org/10.1007/s40430-021-03064-8.

43. Agarwal R, Gupta V, Singh J (2022) A novel drill bit design for reducing bone-chip morphology in orthopaedic bone drilling. In: *Materials Today: Proceedings*. Elsevier Ltd, pp 2–7.

44. Corona PS, Vicente M, Tetsworth K, Glatt V (2018) Preliminary results using patient-specific 3d printed models to improve preoperative planning for correction of post-traumatic tibial deformities with circular frames. *Injury* 49:S51–S59 https://doi.org/10.1016/j.injury.2018.07.017.

45. Mohanavel V, Ashraff Ali KS, Ranganathan K, et al. (2021) The roles and applications of additive manufacturing in the aerospace and automobile sector. *Mater Today Proc* 47:405–409 https://doi.org/10.1016/j.matpr.2021.04.596.

Chapter 6

A case study on the role of additive manufacturing in dentistry

Rahul Jain and Sudhir Kumar Singh
Galgotias Universitys

Rajeev Kumar Upadhyay
Hindustan College of Science & Technology

CONTENTS

6.1	Introduction	100
	6.1.1 Traditional methods used by the dentists in dental treatment	100
	6.1.1.1 Root canal treatment	100
	6.1.1.2 Dental inserts	100
	6.1.1.3 Orthodontics or dental braces	101
6.2	Complications/difficulties in conventional dentistry	102
	6.2.1 Hemorrhages	102
	6.2.2 Neurosensory	102
	6.2.3 Restoration or dental filling	103
	6.2.4 Scaling or dental cleaning	103
6.3	Uses of stereolithographic models in dentistry	104
	6.3.1 Working concept	105
	6.3.2 Clinical uses of rapid prototyping	106
	6.3.3 Use of rapid prototyping in dentistry	106
	6.3.3.1 Orthodontics	106
	6.3.3.2 Oral medical procedure	107
	6.3.3.3 Implantology	107
	6.3.3.4 Maxillofacial prosthesis	107
6.4	Diagnostic methods used by the dentists	108
6.5	Implant design	109
6.6	Material selection for implant	109
6.7	Software used	110
6.8	Cost estimation	110
6.9	Case study	110
6.10	Implantation result	112
6.11	Future scope	113
References		114

DOI: 10.1201/9781003258391-6

6.1 INTRODUCTION

In today's era, Root Canal Treatment is a common procedure in dentistry. In the case of apical periodontitis, endodontic treatment is highly recommended by the medical professionals [1]. Because of the high risk of failure in the treatment of calcified root canals, apicoectomy is another good choice.

Use of additive manufacturing and other related procedures such as guided access procedure and virtual planning are safe and useful techniques that can be used in dentistry to avoid accidents like peroration and deviations. These procedures are also helpful in preserving the tooth structure which is difficult in conventional methods.

In the given case study, with the use of 3D guides and cone beam computed tomography (CBCT) imaging, endodontic treatment is performed in maxillary molar.

6.1.1 Traditional methods used by the dentists in dental treatment

There are a lot of dental strategies that may be blessing to the human beings with amazing regular grins. One has to simply pick the best therapeutic strategy depending on the need.

Some of the most widely used dental treatment techniques are mentioned below:

6.1.1.1 Root canal treatment

This is a dental method to save a normal tooth that has been seriously contaminated and hence decayed. Dental pulps are connective tissues present at the focal point of the tooth containing nerves and lymph tissues. A physical issue or profound pit debilitates the dental mash and causes serious tooth throb [2]. In any case, before the patient simply gets it removed, there should be a discussion with dental specialists as to whether root canal treatment (RCT) is possible. With RCT, one can keep the normal tooth, as treatment can guarantee an easy condition for a day-to-day existence time in the majority of cases.

Root canal treatment involves eliminating the tainted mash, followed by cleaning of the region, and then fixing the tooth with lasting filling. A dental crown is then embedded to safeguard the root canalled tooth.

6.1.1.2 Dental inserts

Any individual who goes through an extraction or has lost a tooth because of injury or ailment isn't bound to live with a missing tooth for the rest of his or her life. A dental insert is a successful swap for the missing tooth.

It is a basic surgery, where an artificial dental embed, usually comprising titanium, is set inside the jaw bone where the tooth is missing [3]. In the end, dental inserts are handled to hold crowns that look precisely like common teeth. It should be noticed that dental implants don't slacken as false teeth. Also, in the same cases, they are perhaps the most mainstream treatment method for related issues [4].

6.1.1.3 Orthodontics or dental braces

Dental supports are the required solutions for individuals with improperly aligned bites, abnormal teeth, jaw problems, and so forth. Crowded teeth are hard to clean and may prompt different oral issues. Subsequently, to resolve such issues, selecting the desired support is a decent alternative. One can discover data about dental support treatments in India across different websites. Dental supports are normally actualized at a more youthful age. More than dental crises, which require a prompt consideration and the board, the event of "confusions" is of higher occurrence in dental practice [5]. The confusions might be quick or postponed and are identified with patient's resilience level, materials utilized, and treatment methodology.

In an interdisciplinary dental practice, the most well-known difficulty is aspiration. Desire may be for general dental replacement or for a cracked section, a minor augmentation acrylic removable prosthesis, crowns during expulsion, instrument slippage, particularly reamers or documents. Aspiration causes aviation route block, which is shown as the all-inclusive sign of "choking." Expulsion of broken instruments is performed utilizing ultrasonics, working magnifying lens, or microtube delivery techniques.

Sensitivity is another complication regularly experienced by dental trained professionals.

Sensitivity can be to latex, mercury, elastic dam, and impression material. Appearances of hypersensitivity incorporate pruritus, erythema, utricaria, and angioneurotic edema. Limiting latex openness is the best option while treating latex-delicate patients. Latex options (vinyl, nitrite, or silicone) and sans powder gloves ought to be utilized to forestall sharpening. Fixers like formacresol and devitalizers to be utilized cautiously to forestall synthetic consumes [6]. Hypersensitivity to compounds like nickel-chromium and chromium-cobalt has likewise been experienced.

Complexities including local anesthetics are touchiness, harmful responses, and hypersensitivity. The most extreme type of hypersensitivity is anaphylaxis which is a hazardous summed up or fundamental response. Anaphylaxis may be either allergic or non-allergic. Unfavorably susceptible hypersensitivity can be quick because of IgE or deferred which is lymphocyte interceded.

6.2 COMPLICATIONS/DIFFICULTIES IN CONVENTIONAL DENTISTRY

Entanglements can be identified either with the medical procedure or implant arrangement. The intra-employable difficulties identified with a medical procedure are hemorrhages, neurosensory change, harm to the adjoining teeth, and mandibular cracks.

6.2.1 Hemorrhages

Hemorrhages in the mandible most often happen in the intra-foraminal locale by harm to the plummeting palatine corridor or the back palatine vein. Respiratory block has likewise been accounted for because of the hole of the conduits providing the mandible. This is accepted to be because of the enormous interior drain caused due to the vascular injury in the floor of the mouth which makes an expanding, delivering bulge, and dislodging of the tongue, thereby resulting in the obstruction of the airways. Hemorrhages can be controlled with firm finger pressure for the purpose of draining, but if compressions do not stop the bleeding, anastomoses may require ligation.

6.2.2 Neurosensory

Another problem identified with the medical procedure is neurosensory unsettling influence which manifests as sedation, paresthesia, hypoesthesia, or dysesthesia. On the off chance that the patient experiences paresthesia yet the implant is set accurately with no harm to the nerve, at that point recovery of the implant isn't exhorted; rather, the patient is advised to sit tight for recuperation. Nonetheless, if the nerve is being packed, it is consistently prudent to eliminate the implant to stay away from perpetual neural harm. Harm to the contiguous teeth happens because of the absence of parallelism of the implant with the neighboring teeth [7]. Subsequently, it is consistently required that a distance of 1.5 mm should be kept up from the adjoining teeth.

If there is the possibility of damage, treatment of the influenced teeth incorporates endodontic treatment, periapical medical procedure, apicectomy, or extraction. Mandibular breaks are uncommon and happen when implants are set in atrophic mandible.

Intricacy related with implant situation in particular includes loss of essential steadiness which can be ascribed to the extreme working behavior of the implant bed, low-quality bone, or the application of short implants. Increase in temperature because of the inordinate speed of the drill produces fibrosis, osteolytic degeneration, necrosis, and increased osteoclastic action. Loss of essential steadiness can be overseen by using a more extensive and longer self-tapping implant. Another conceivable difficulty is the indication of dehiscence or fenestration, overseeing which includes filling the bone imperfection with bone grafts and resorbable or non-resorbable membranes [8]. During the insert position in the maxilla in zones near the sinus or during sinus lift methods, difficulties including break of the Schneider membrane can take place. Contingent upon the width of the tear, a resorbable film might be used which serves to contain the bone join material, or in the event that the tear is extremely wide, at that point a medical procedure is deferred.

Another complexity is the uprooting of the inserts into the maxillary sinus during the medical procedure or in the postoperative period. At times, it can prompt sinusitis or even stay asymptomatic.

These confusions and emergencies can be limited by fitting pre-surgical arranging, use of precise careful strategies, postsurgical follow-up regarding the osseointegration time frame, proper plan of the superstructure, biomechanics, and supporting fastidious cleanliness during the upkeep phase [9]. However, today, oral remedies utilizing supports are additionally accessible for grown-ups.

6.2.3 Restoration or dental filling

This is perhaps a well-known and basic dental treatment in India to treat a rotten tooth. Nonetheless, this treatment is chosen when the cavity has not arrived at the focal piece of the tooth known as pulp. This technique includes eliminating the rotten piece of the tooth and afterward cleaning the area to make it liberated from any bacterial invasion [10]. The zone is then loaded up with filling material, for example, porcelain, combination of filling materials, or in some cases gold. A decent dental filling supports for quite a long while if legitimate oral cleanliness is followed. On the off chance that a filled tooth isn't taken appropriate care of, extreme pit may build up, which may require root canal or extraction.

6.2.4 Scaling or dental cleaning

It is a process that targets eliminating stains, tartar, and plaque aggregated on the teeth, consequently giving shining white teeth. Normal brushing and flossing may not guarantee precious perfectly clean teeth and fresh breath. Accordingly, dental hygienists propose teeth cleaning after each 8–12 months. Tartar developed may prompt extreme gum sicknesses and periodontal issues. In this manner, intermittent visits to the dental specialist for routine teeth cleaning are essential to keep these illnesses away [11]. Dental cleaning is followed by bleaching for individuals who desire genuine white teeth. In any case, this is entirely optional and completed as a restorative methodology rather than an oral health necessity.

6.3 USES OF STEREOLITHOGRAPHIC MODELS IN DENTISTRY

The term rapid prototyping (RP) assigns a bunch of advancements that permits the acknowledgment of programmed actual models dependent on plan information, all with the assistance of a computer. These "three-dimensional printers" permit architects to rapidly produce characterized models of their plans, as opposed to the basic two-dimensional pictures [12]. These models of such accomplishments give important visual guidance. The move from the visual to the visual-material portrayal of actual items presented another sort of communication called the "contact to understand." In the beginning of RP, auto and aviation businesses overwhelmed the RP application. Yet, this is not true anymore as RP has spread into numerous different enterprises.

It has altered science and design by integrating itself into numerous aspects of contemporary life, from diversion to medication. Everything began during the 70s, when it spread another strategy for clinical data–based X-beam, that is, the tomographic assessment or computerized tomography (CT). RP advances are another methodology for careful arranging and reenactment. They recreated anatomical items as three-dimensional actual models, which give the specialist a practical impression of complex constructions before a careful intercession [13,14].

The need for confronting the mathematical intricacy has brought RP into the dental field. It can possibly turn into the cutting edge in creation techniques in dentistry. Its past realized commitment related with the conclusion, training, and careful arranging. Including prosthodontics, this innovation is being utilized in diverse zones of dentistry.

The development of the RP innovation into prosthodontics has enhanced the lab and clinical techniques by dispensing with or annulling some middle-of-the-road stages and making independent the nature of the results from the professionals' abilities. This demonstrates the capability of the new strategy, which is equipped for supplanting the customary "impression-taking and waxing" technique. RP techniques are utilized to generously abbreviate the ideal opportunity for creating examples, molds, and models. There are various RP advances accessible.

In any case, the field of Rapid Prototyping is still young and requires a lot of effort to develop precision, dependability of the framework, and speed while also broadening the scope of materials for model development. So, the dentists ought to know about likely areas for errors inside models and survey the source picture in situations where models' reliability is uncertain [15]. Another zone of progress will be the expense effectiveness, as most RP frameworks are as of now too costly to ever be moderate.

6.3.1 Working concept

Flow chart describing classification of RP method is shown in Figure 6.1. This new RP innovation is based on the deterioration of 3-D computer models in the layers segment cross over meagre, followed by genuinely framing layers and heaping layer over layer. The age of 3D articles thus is a thought dating back to the time of the inception of human civilization. The advancements since the Egyptian pyramids were presumably block created layer over layer [16, 17]. The RP process chain is shown in Figure 6.2.

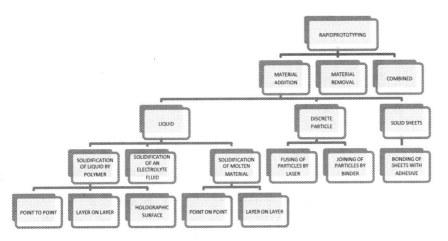

Figure 6.1 Flow chart describing classification of the rapid prototyping method.

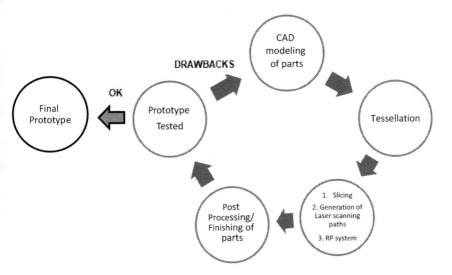

Figure 6.2 Rapid prototyping process chain.

6.3.2 Clinical uses of rapid prototyping

Clinical RP is characterized as the assembling of dimensionally precise actual human life system models got from clinical picture information using an assortment of RP technologies. Some RP machines had just been in exploratory use during the 1970s, and CT was concocted during the 1960s by Godfrey N Hounsfield, a hardware engineer, in a joint effort with Allan McLeod Cormack, a physicist. In any case, it was distinctly during the 1990s that a real three-dimensional model was worked to recreate the life structures of a patient dependent on CT pictures got during that patient's assessment, on account of advances in CT scanner quality and the improvement of explicit programming for this reason [18]. RP has been applied to a scope of clinical strengths, including oral and maxillofacial medical procedure, neurosurgery, dental implantology, and muscular health.

Difficult illnesses in medication regularly request tedious medical procedure. Careful arranging attempts are taken to limit the length of medical procedure to lessen the danger of difficulties. Notwithstanding, the ordinarily utilized imaging modalities, 3D perception methods can be applied for supporting the arranging cycle. Such visual portrayal of clinical articles permits reenactment of surgeries before medical procedure [19]. The best preferred position of RP advances is the exact proliferation of articles from a three-dimensional clinical picture informational collection as an actual model which can be taken a gander (at)X and contacted by the surgeon.

Clinical RP is likewise being created for use in dental implants. Furthermore, precision was achieved through the use of quick prototyped careful aides for making osteotomies in the jaw, and a computerised helped plan or helped make (CAD/CAM) method for dealing with the manufacture of incomplete dental structures was introduced.

6.3.3 Use of rapid prototyping in dentistry

6.3.3.1 Orthodontics

Utilizing cutting edge CAD/CAM innovation, the two typically separate cycles of section creation and section situating are combined into one unit. In this cycle, the interest for most extreme distinction with, at the same time, limited space necessities are put reliably into use. One more inventive use of this innovation was to make an overcrown ready to open the bite through clinical crown protracting of the mandibular second premolars. Some innovation gives clear plastic orthodontic treatment gadgets [20]. Every 14 days, the patient receives another arrangement of assistance, such as aligners, which are expected to keep the teeth moving. This innovation uses a few stereolithography machines to create models whereupon plastic sheets are shaped. Informational collections are gotten by digitizing an impression taken of the patient's teeth. The subsequent point sets are isolated into singular tooth calculations, which are then situated by the orthodontist's treatment plan.

6.3.3.2 Oral medical procedure

Anatomic clinical models worked with RP innovations address another methodology for careful arranging and recreation. These methods permit one to duplicate anatomical articles, for example, 3D actual models of the skull or different constructions, which give the specialist a sensible impression of complex designs before careful mediation [21]. The move from the visual to the visual-material portrayal of anatomical items presents another sort of collaboration called "contact to comprehend." Clinical information demonstrates that computer supported RP might be of an incentive in limiting the extraoral time and conceivable injury to relocated teeth during the cycle of autotransplantation.

6.3.3.3 Implantology

Since the approach of osseointegration, the usage of dental implants has advanced quickly in the course of the most recent decade. Exploration of oral implantology has prompted refinements bringing exceptionally fruitful and unsurprising therapeutic alternatives for somewhat just as totally edentulous patients; in any case, ill-advised embed situation can have a significant and frequently impeding impact on the drawn out consistency and achievement of the embed upheld prosthesis [22,23].

The use of computer-aided design or assisted manufacturing (CAD/CAM) innovation has grown in popularity in implant treatment. The applications relate to 3D imaging, 3D programming for treatment arranging, creation of computer produced careful aides utilizing added substance RP, just as manufacture of all-clay rebuilding efforts uses subtractive RP. RP innovation considers mechanical manufacture of redid three-dimensional articles from computer helped plan data.

6.3.3.4 Maxillofacial prosthesis

Non-appearance of all or a piece of the outer ear might be either obtained or inborn. When endeavoring to reestablish this part with prosthesis, the prosthesis ought to preferably be modified to reestablish the life systems as intently as could really be expected. In this manner, it very well might be useful to have the earlier information on normal qualities for each list and utilize these qualities to help develop prosthesis of the suitable size and shape [24]. Nonetheless, singular extent records a shift from the normal, so where the imperfection is one-sided it is more pragmatic to think about and copy extents from the no defect side. This cycle can be troublesome and tedious and requests a significant degree of imaginative ability to shape an identical representation and accomplish a decent tasteful match. Likewise, patients with existing prostheses may require continuous substitutions in light of shading changes, tearing, loss of fit, maturing, tainting of the material, and normal wear. Ordinary duplication systems are regularly untrustworthy and off base, as mistakes may happen at any of the numerous stages during manufacturing [25].

The coming of CT and attractive reverberation imaging with 3D portrayal of human body structures has cleared new points of view for plan and creation in the clinical field computer control of the information which takes into consideration reflecting or changes to build up the specific measurements required, and a computer numeric controlled (CNC) processing machine can be used to produce a format for the last prosthesis [26]. CNC processing, notwithstanding, is restricted by challenges experienced when attempting to repeat the intricate life structures of inside highlights.

The improvement of RP industries has prompted the formation of redid 3D anatomic models that show a degree of unpredictability obscure with CNC-based gear, essentially on the grounds that RP strategies utilize an added substance interaction of building an article of various layers characterized by a computerized model that has been for all intents and purposes sliced. This takes into account the creation of complex shapes with inner details and other areas. One such process is stereolithography, which produces three-dimensional items by relieving a fluid gum under a computerguided laser. The upside of such a framework is the capacity to project straightforwardly from a wax model.

6.4 DIAGNOSTIC METHODS USED BY THE DENTISTS

There are various diagnostic methods that are used by the dentists nowadays. These techniques are broadly classified into the below mentioned categories:

  a. Analog and digital
  b. Intraoral and extraoral
  c. Ionizing and non-ionizing imaging
  d. Two-dimensional and three-dimensional imaging

Among the list of so many diagnostic techniques, computed tomography is in great demand. CBCT and tuned aperture computed tomography (TACT) are the much in demand imaging techniques in today's market.

In this study, CBCT is used because of the following key benefits:

- It improves the X-ray utilization and requires comparatively low energy as compared to other techniques.
- No extra mechanism is required in moving the patient during acquisition.
- High image quality and high resolution makes it easy to detect a variety of infections, tumors, cysts, traumatic injuries, and developmental anomalies involving the maxilla-facial structures.

- As compared to helical CT, CBCT has high image quality for the highest resolution modalities.
- For a perfect study of the relationship of the adjacent structures, CBCT provides high and clear spatial resolution of bone and teeth.

6.5 IMPLANT DESIGN

While designing the implant, the following design considerations were taken into account:

- Implant diameter
- Implant length
- Implant shape

6.6 MATERIAL SELECTION FOR IMPLANT

There are various types of materials available in the market for making the implants. Some of them are as follows:

a. Metals
  - Titanium
  - Titanium alloy
  - Co-Cr alloy
  - Stainless Steel
  - Tantalum
b. Ceramics
  - Hydroxyapatite
  - Alumina
  - Carbon-Silicon
  - Beta-Tricalcium Phospate
  - Bio-glass
c. Polymers
  - Polymethylmethacrylate
  - Polytetrafluoroethylene
  - Polyethylene
  - Polysulphone
  - Polyurethane

Among a wide range of materials available for the implants, PLA (polylactic acid) is mainly used in the printing of dental implants prototypes. PLA is easily available and is a thermoplastic extracted from the corn starch or sugarcane that is not at all harmful for the human body.

6.7 SOFTWARE USED

Using a structured light 3D scanner, the patient's physical models of dental anatomy were scanned and various diagnostic models like CBCT were used. Using the data, collected from CBCT, the model of maxilla was generated in .stl format and later converted into .iges format and used in Catia V5 software. This 3D model helped in evaluating the size of implants, orientation, and best position. Surface modeling along with simple Boolean operation was used to design the drill guide for the implant positioning in the part design Catia V5 module.

6.8 COST ESTIMATION

Usually in India, dental implants cost between 35,000 INR and 45,000 INR. The material used for the implant, type of tooth implant, design of the implant, and many more reasons decide the cost of implantation. In this case study, the final cost is about 50,000 INR.

6.9 CASE STUDY

For conducting the case study, 10 renowned dental clinics were visited to know the various dental problems and their causes.

After the visit, 143 patients with different dental problems were interviewed and the major problem found was that of tooth decay. The major cause of this problem is bad eating habits and lack of oral hygiene.

After knowing the problem and the root cause of that problem, several brain storming sessions were carried out with dental experts to decide the best suited treatment and incorporation of additive manufacturing in dentistry.

Among those ten dental clinics, Surya Dental Clinic, Agra (Uttar Pradesh, India) was finalized for the study.

A case study was carried out on 52 persons belonging to different age groups and facing different dental problems. Table 6.1 shows the data collected from the clinic for the case study:

Table 6.1 Data collected from clinic for the case study

S. no.	Age group	No. of patients	Problem diagnosed in maximum cases
1	15–24	04	Tooth wear & malocclusion
2	25–34	11	Dental caries & periodontal problems
3	35–44	16	Dental caries & tooth wear
4	45–54	12	Tooth wear & periodontal problems
5	55–64	06	Tooth wear & periodontal problems
6	65 and above	03	Tooth wear & periodontal problems

Depending upon the available resources and patient consent, this study was performed. A person facing a problem in the left upper molar reported for treatment. The dentist was unable to explain the pulp reaction to heat and other examination in the second maxillary molar during the usual check-up, while a mild sensitivity to percussion was also seen. Extensive periapical bone resorption and large-scale restoration in tooth were confirmed, following the clinical examination and CBCT scan. The operator was unable to locate the entrance of the palatal canal of the second molar, safely and correctly, proving the calcification results from the original CBCT imaging.

To obtain the periapical area's thorough view, a CBCT imaging unit of high quality was used. The pictures got from CBCT examining demonstrated a radiolucent zone of 5.64 mm. The root canal must be found in the second molar's middle third of the palatal root. Thinking about the issue, guided access method in tooth was recommended to the patient. A 3D-printed plan was used to get to the root canal lumen.

In order to make the guide, various steps were taken; first of all, intraoral impression was collected, and then this impression was scanned using the desktop, R700. The scanned impression was uploaded into virtual implant planning software. After performing this, additional CBCT images were uploaded. The visible structures based on radiography, the CBCT scan, and surface scan were matched. Subsequent to arranging the drill position, a virtual layout was planned, using the SimPlant programming format originator device. A controlling sleeve was modified for the drill by methods for a product device and essentially joined into the arranging library prior to making the format. Obsession sleeves were additionally made for the reason of settling the guide, forestalling the hole drill from veering off from its direction made dependent on the tomographic arranging.

The virtual layout was traded as an standard triangle language (STL) document and transferred to a three-dimensional printer. The recently referenced sleeve was incorporated into the printed format to control the drill during cavity arrangement.

To make the surgery comfortable and less painful, local anesthesia (a common procedure) was administered to the patient; after this, bone cutting was performed in two focuses under steady water system with saline water for the reason of focusing the guide in a proper way. From that point onward, guided permission was gained through the calcified root portion utilizing a comparable drill at a speed of 1200 rpm. At the point when the guide was dispensed with, the tooth was confined using a rubber dam to affirm whether the channels had been gotten to. At the same time, it was possible to affirm the patency length of the canal (the palatal canal of the upper second molar) by joining the apical locator with the radiographic assessment. Instrumentation of the canal was performed using the WaveOne Gold Large responding instrument with 2.5% sodium hypochlorite as the irrigant. In the wake of drying the canals, calcium hydroxide P.A. and glycerin were used as intracanal remedies in treatment gatherings.

112 Additive Manufacturing

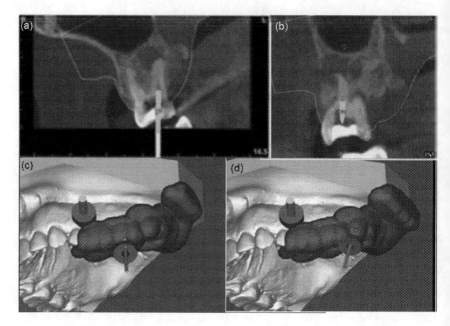

Figure 6.3 Demo of the design of virtual drill using SimPlant software [21].

Demo of the design of the virtual drill using SimPlant software is shown in Figure 6.3. A composite resin containing filler was used as a temporary filling material. Following 14 days, the canal was filled by the hydraulic compression process using FM gutta-percha cones and an epoxy-based cement as a sealer was used. The gutta-percha cones were adjusted at 0.5 mm short of the patency length of the canal. Accordingly, the access cavities were cleaned, and the tooth was reestablished.

Following the filling of the root canals of the maxillary second molar, the patient neither introduced pain manifestations in the area nor reacted decisively to the percussion tests. A quarter of a year after finishing up the case, relapse of the sores was noticed. The 1-year follow-up uncovered a radical decrease of the periradicular injuries and torment indications and a negative reaction to the percussion tests.

6.10 IMPLANTATION RESULT

The number of senior citizens needing endodontic treatment is increasing. These patients may have teeth with partially or completely calcified pulp chambers and obstructed root canals as a result of dentin relationship for the rest of their lives.. Youth patients who may show up with incompletely or totally calcified root canals because of dental injury are another significant

Role of additive manufacturing in dentistry **113**

classification. Fifteen percent of damaged incisors present incomplete pulp annihilation; 1% turned out to be completely calcified, which could be viewed as an indication of pulp recuperating, and accordingly a case in which endodontic treatment isn't required. All things considered, there is a 1%–27% danger of these pulps getting necrotic eventually.

The writing shows that achieving a satisfactory primary opening furthermore, distinguishing the passage orifice of the canals is very testing when considering calcified cases added to pulp disease and ordinarily prompts a radical loss of dental construction related with a more prominent danger of fracture and increased failure rate.

CBCT imaging has ended up being an entirely significant tool in endodontics, assisting with the finding of resorptions, periapical pathologies, calcifications, and root morphology, among others. Moreover, this method has added to expanding the achievement pace of endodontic medicines by advancing specialized treatment arranging. It is also extensively used in dental medical procedure as a planning aid as well as a manual for bone penetrating prior to implant arrangement.

Enlivened by the guided implant method, the guided endodontic procedure comprises getting to and finding root canals with extreme calcification by methods for managing formats made by tomographic arranging as revealed by certain authors. This strategy is by all accounts a safe and clinically achievable strategy, particularly when the calcified canals couldn't be gotten to by regular endodontic procedures in an unsurprising way. The technique lessens the treatment time; what's more, it can be performed by a less experienced operator. The authors were consistent in recognizing the precision of the method and its significance for the additional opportunity of haggling seriously calcified root canals. Accordingly, even without the guide of the operating microscope, the guided endodontic treatment could be very supportive to experts when confronting more intricate cases. The guided template sleeves direct the situation of the access burrs, producing more prominent accuracy imitating sufficient tomographic planning.

6.11 FUTURE SCOPE

The guided access process for the endodontic treatment of calcified upper molars was demonstrated to be protected and precise, encouraging access and endodontic treatment in general under safe, speeded up, and unsurprising conditions. CBCT imaging and the access guides were basic and key segments for playing out this new method. The effortlessness of this method permits it to be performed even by less experienced experts. The advancement of more sufficient drills in both length and gauge specifically planned and more fit for meeting the assumptions and prerequisites for use in endodontics is a genuine requirement and should be tended to in future works.

REFERENCES

1. American Association of Endodontists (2006) Endodontic case difficulty assessment form and guidelines. [WWW document] URL https://www.aae. org/uploadedfiles/dental_professionals/endodontic_case_assessment/2006 casedifficultyassessmentformb_edited2010.pdf [accessed on 22 February 2017].
2. Agamy HA, Bakry NS, Mounir MM, Avery DR (2004) Comparison of mineral trioxide aggregate and formocresol as pulp-capping agents in pulpotomized primary teeth. *Pediatric Dentistry* 26, 302–9.
3. Andreasen JO (1970) Luxation of permanent teeth due to trauma. A clinical and radiographic follow-up study of 189 injured teeth. *Scandinavian Journal of Dental Research* 78, 273–86.
4. Andreasen FM, Zhijie Y, Thomsen BL, Andersen PK (1987) Occurrence of pulp canal obliteration after luxation injuries in the permanent dentition. *Endodontics & Dental Traumatology* 3, 103–15.
5. Brodin P, Linge L, Aars H (1996) Instant assessment of pulpal blood flow after orthodontic force application. *Journal of Orofacial Orthopedics* 57, 306–9.
6. Capar ID, Uysal B, Ok E, Arslan H (2015) Effect of the size of the apical enlargement with rotary instruments, single-cone filling, post space preparation with drills, fiber post removal, and root canal filling removal on apical crack initiation and propagation. *Journal of Endodontics* 41, 253–6.
7. Connert T, Zehnder MS, Weiger R, Kühl S, Krastl G (2017). Microguided endodontics: accuracy of a miniaturized technique for apically extended access cavity preparation in anterior teeth. *Journal of Endodontics* 43, 787–90.
8. Cvek M, Granath L, Lundberg M (1982) Failures and healing in endodontically treated non-vital anterior teeth with posttraumatically reduced pulpal lumen. *Acta Odontologica Scandinavica* 40, 223–8.
9. Delivanis HP, Sauer GJ (1982) Incidence of canal calcification in the orthodontic patient. *American Journal of Orthodontics* 82, 58–61.
10. European Society of Endodontology (2014) European Society of Endodontology position statement: the use of CBCT in endodontics. *International Endodontic Journal* 47, 502–4.
11. Fleig S, Attin T, Jungbluth H (2016) Narrowing of the radicular pulp space in coronally restored teeth. *Clinical Oral Investigations.* doi:10.1007/s00784-016-1899-8 [Epub ahead of print].
12. Holcomb JB, Gregory WB Jr (1967) Calcific metamorphosis of the pulp: its incidence and treatment. *Oral Surgery, Oral Medicine, Oral Pathology* 24, 825–30.
13. Hussey DL, Biagioni PA, McCullagh JJ, Lamey PJ (1997) Thermographic assessment of heat generated on the root surface during post space preparation. *International Endodontic Journal* 30, 187–90.
14. Jacobsen I, Kerekes K (1977) Long-term prognosis of traumatized permanent anterior teeth showing calcifying processes in the pulp cavity. *Scandinavian Journal of Dental Research* 85, 588–98.
15. Johnstone M, Parashos P (2015) Endodontics and the ageing patient. *Australian Dental Journal* 60(Suppl. 1), 20–7.

16. Kiefner P, Connert T, ElAyouti A, Weiger R (2016) Treatment of calcified root canals in elderly people: a clinical study about the accessibility, the time needed and the outcome with a three year follow-up. *Gerodontology.* doi:10.1111/ger.12238 [Epub ahead of print].

17. Krastl G, Zehnder MS, Connert T, Weiger R, Kuhl S (2016) Guided Endodontics: a novel treatment approach for teeth with pulp canal calcification and apical pathology. *Dental Traumatology* 32, 240–6.

18. Lang H, Korkmaz Y, Schneider K, Raab WH (2006) Impact of endodontic treatments on the rigidity of the root. *Journal of Dental Research* 85, 364–8.

19. Mai HN, Lee KB, Lee DH (2017) Fit of interim crowns fabricated using photopolymer-jetting 3D printing. *The Journal of Prosthetic Dentistry* 118(2), 208–15.

20. Connert T, Zehnder MS, Amato M, Weiger R, Kühl S, Krastl G (2018) Microguided Endodontics: a method to achieve minimally invasive access cavity preparation and root canal location in mandibular incisors using a novel computer-guided technique. *International Endodontic Journal* 51(2), 247–55.

21. Sônia TD, Camila de Freitas MB, Santa-Rosa CC, Machado VC (2018) Guided endodontic access in maxillary molars using cone-beam computed tomography and computer-aided design/Computer-aided manufacturing system: a case report. *Journal of Endodontics* 44(5), 875–9.

22. Ballard DH, Trace AP, Ali S, Hodgdon T, Zygmont ME, DeBenedectis CM, Smith SE, Richardson ML, Patel MJ, Decker SJ, Lenchik L (2017) Clinical applications of 3D printing: primer for radiologists. *Academic Radiology.*

23. Anadioti E, Kane B, Soulas E (2018) Current and emerging applications of 3D printing in restorative dentistry. *Current Oral Health Reports*, 1–7.

24. Osman RB, van der Veen AJ, Huiberts D, Wismeijer D, Alharbi N (2017) 3D-printing zirconia implants; a dream or a reality? An in-vitro study evaluating the dimensional accuracy, surface topography and mechanical properties of printed zirconia implant and discs. *Journal of the Mechanical Behavior of Biomedical Materials* 75, 521–8.

25. Zaharia C, Gabor AG, Gavrilovici A, Stan AT, Idorasi L, Sinescu C, Negruţiu ML (2017) Digital dentistry—3D printing applications. *Journal of Interdisciplinary Medicine* 2(1), 50–3.

26. Sato M (2020) Efficacy of 3D-printed guide vs conventional method for conservative endodontic access preparation in extracted upper molars. Master's Theses, 1523.

Chapter 7

Role of additive manufacturing in biomedical application

Vipin Goyal, Ajit Kumar, and Girish Chandra Verma
Indian Institute of Technology (Indore)

CONTENTS

7.1	Introduction	118
7.2	Classification of the additive manufacturing process	118
7.3	Biomaterials	120
7.4	Applications of additive manufacturing biomedical domain	122
	7.4.1 Tissue engineering	122
	7.4.2 Tissue regeneration	122
	7.4.3 Implants	122
	7.4.3.1 Orthopedics	123
	7.4.3.2 Dentistry	123
	7.4.4 Pharmaceuticals	123
	7.4.5 Surgical tools	124
	7.4.6 Operative planning	124
	7.4.7 Implant tissue interface	124
	7.4.8 Personalized protective equipment during COVID 19	124
7.5	Advantages of additive manufacturing over conventional manufacturing in the biomedical field	125
7.6	Limitations of additive manufacturing in the biomedical domain	126
	7.6.1 Cost-effective only for low-volume production	126
	7.6.2 Limited material option	126
	7.6.3 Poor mechanical properties	126
	7.6.4 Low-dimensional accuracy	126
7.7	Mechanical properties of biomedical parts	126
	7.7.1 Tensile and compressive strength	127
	7.7.2 Fracture toughness	127
7.8	Future aspects	127
7.9	Conclusion	128
References		128

DOI: 10.1201/9781003258391-7

118 Additive Manufacturing

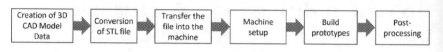

Figure 7.1 Steps involved in the fabrication of medical components using additive manufacturing (1).

7.1 INTRODUCTION

The conventional manufacturing technique is a subtractive machining technique involving several steps and requires a huge amount of time and money to produce components. This traditional technique does not allow the production of complex and intricate parts necessary for biomedical application. However, in recent years, 3D printing, also known as additive manufacturing (AM), has evolved as a reliable technology for generating complicated and patient-specific medical components using biomaterial powder and computer-aided design. As per ASTM F-2792, AM is a material joining process in a layer upon layer manner using 3D model data. Steps used to produce parts using an AM process are depicted in Figure 7.1.

With this technology, a complex and robust design can be produced that is nearly impossible to produce using the conventional manufacturing technique. Consequently, its use in the biomedical industry is increasing rapidly due to its applications in tissue engineering and tissue generation, surgical implants, surgical planning, and pharmaceuticals since it allows for the printing of customized medical parts at a lower cost (2–7).

Many AM methods have been developed since 1980 for a variety of biomaterials, including polymers, metals, bioceramics, and a combination of two or more of these for obtaining desired mechanical and biological characteristics such as biocompatibility, strength, corrosion, and wear resistance, and reduced shielding stress (8).

Medical and engineering professionals can use this approach to complement each other to plan surgical procedures, design implants, deliver therapeutics, test drugs, model diseases, and fabricate soft and hard tissues (9). This chapter discusses the advantages, limitations, mechanical properties, and build processes of AM for various biomedical applications. In addition, this chapter outlines current research gaps and future directions of the AM technique for biomedical applications.

7.2 CLASSIFICATION OF THE ADDITIVE MANUFACTURING PROCESS

Many AM processes, such as vat polymerization, material extrusion (ME), material jetting (MJ), sheet lamination, direct energy deposition (DED), and powder bed fusion (PBF), have been established, according to the ISO/

ASTM52900 standards. Figure 7.2 depicts the wide categories of AM. Despite the fact that several sophisticated AM processes are available, such as selective laser sintering and selective laser melting, the advantages of selective laser sintering and selective laser melting include high accuracy, quick manufacturing, and a greater degree of design flexibility.

Table 7.1 summarizes the advantages, limitations, and applications of various AM techniques, definitions, and biomaterials. The following is a quick summary of each process.

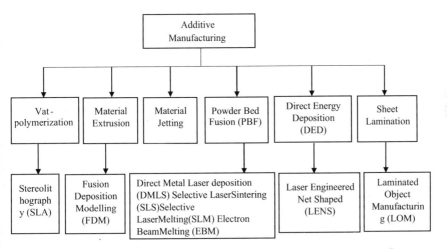

Figure 7.2 Classification of the additive manufacturing technique (10).

Table 7.1 The additive manufacturing process with a brief description, pros, cons, and applications (11)

Process	Brief description	Pros	Cons	Application
Vat polymerization (SLA)	Liquid photopolymer in a vat is selectively cured by UV light	Better resolution, lowest stress distribution	Low strength and only polymers can be used	Tissue engineering and regenerative medicine, dentistry, surgical model, medical devices
Material extrusion	Material is selectively dispensed through a nozzle and deposited using thermal and cross-linking	No constraints of material, quick design change, lightweight geometries, design for customization (12)	No choice of a variety of materials	Medical aids, supportive guides spline, prosthesis

(Continued)

120 Additive Manufacturing

Table 7.1 (Continued) The additive manufacturing process with a brief description, pros, cons, and applications (11)

Process	Brief description	Pros	Cons	Application
Material jetting	Droplets of build material are selectively deposited using UV light	Quick design change, lightweight geometries	Limited choice of materials	Medical models and bio-manufacturing
Sheet lamination (LOM)	Sheets of material are bonded and form an object	Low manufacturing cost for high volume	Poor surface quality and complicated shape could not be produced	Biomedical tools, instruments, and medical devices
Powder bed fusion (DMLS, SLM, EBM)	The laser energy melts or fuse the metal powder	Better mechanical properties	Surface quality is so poor, high inaccuracy, post-processing is required, and costly process	Implants, bone fixation devices, artificial joints tissue engineering, and generation
Direct energy deposition	The powder material is fused or melted by thermal or laser energy	High density, better mechanical properties, high deposition rate, less feedstock wastage	Poor surface finish, poor resolution, expensive machine setup	Orthopedic implants, bone fixation, artificial joints

7.3 BIOMATERIALS

The most common use of AM in the biomedical field is to produce patient-specific components that must function properly. The essential necessity of a biomedical component is improved mechanical properties, which may be achieved by good material selection. Biomaterials should have properties like being easy to print, biocompatible, corrosion-resistant, having a high strength to weight ratio, and being non-toxic, among others. Biopolymers, bio-metals, and bioceramics are the three main forms of bio-materials employed. Metals are often employed in high-strength implants and fixation systems (13). Table 7.2 compares various applications and AM techniques with their respective applications.

Table 7.2 Biomaterials with their application and processing method

Process	Materials used	Application	Advantages	Ref.
Polymer	Polycaprolactone (PCL)	Bone tissue engineering scaffold	Biocompatible, biodegradable, proliferation, differentiation, high temp stable, faster cell adhesion	(14–16)
	Polylactic co-glycolic acid (PLGA)	Scaffold, graft material, fixation devices, tissue reconstruction	Biodegradable, biocompatible, create porosity	(17)
Metals	SS316	Orthopedic implants, fixation devices	Easily available	(18)
	Ti alloy	Orthopedic implants, fixation devices	Highly corrosion resistance, lightweight	(19)
	Co-Cr-Mo	Femoral knee implants	High wear resistance	(20)
	Mg	Implants, fixation devices, ligatures, and wire	Easily available, lightweight, low cost	(21)
Ceramic	Tricalcium phosphate	Tissue engineering	Biocompatible, bioactive, osteoinductive	(22)
	Hydroxyapatite	Tissue engineering, scaffold, bone filler, drug delivery, implant coating	Biodegradable, biocompatible	(23)
	Bioglass	Tissue engineering, bone defect fillers	Bioactive, osteogenesis	(24,25)
Composites	HA+ PCL	Bone tissue engineering scaffold	Bioactive and excellent thermal stability	(26)
	Bioglass+ PCl	Scaffold	Excellent cell adhesion, proliferation, cell adhesion	(27)

7.4 APPLICATIONS OF ADDITIVE MANUFACTURING BIOMEDICAL DOMAIN

Although AM was initially employed in the medical field in 1990, it has provided a breakthrough platform for the development of biomedical applications in recent decades. Tissue engineering and tissue generation are the first two types of applications for AM employing biomaterials in biomedical applications, in which cells require 3D support with apertures for adhering, growing, and differentiating into functional tissues or organs (28,29). Implants, such as orthopedic and dental implants, are another area of use. These implants are custom-made for each patient. Personalized protective equipment (PPE), including face shields and hand protection equipment, was manufactured by AM during the COVID 19 epidemic (30). This section covers the application of AM in a variety of biomedical fields, including tissue engineering and production, implants, medicines, surgical instruments, and personal protective equipment (PPE).

7.4.1 Tissue engineering

Tissue engineering serves as a link between biomedical and engineering fields. In the event of an injury or accident, the major goal of this technique is to construct and replace the natural implant with an artificial implant. This technique has to do with live cells, organs, and the treatment of burns. The field of Tissue engineering is revolutionized by AM. The porous structure is essential for bone regeneration, which stimulates fast cell proliferation within the body (31). The fundamental benefit of AM in tissue engineering is that it can produce 3D complicated shapes of organs and body components with great precision and accuracy.

7.4.2 Tissue regeneration

AM is an effective tool for tissue generation in vivo and in vitro studies. AM offers potential implant solutions that could promote both vascularization and bone regeneration. One of the most essential features that can improve osseointegration is porosity, which can be accomplished utilizing various biodegradable bioceramics or biopolymers to create a porous scaffold. Many bioceramics, including HAP, PCL, Mg, and TCP, can be employed for the porous structure.

7.4.3 Implants

In recent years, surgical implants and prostheses have been continuously developing because they exhibit excellent adhesion to the surrounding tissue and increase the functionality of the body parts. The design of implants and prostheses requires a multidisciplinary approach such as materials science, engineering design, biomechanics, molecular biology, bioscience, pharmaceutical, and continuous clinical monitoring. AM is able to provide better osteoinductivity and osteoconductivity that promote bone regeneration and integration of the implant with the surrounding tissue. All of these characteristics are critical in biological applications (32).

7.4.3.1 Orthopedics

In the event of an injury or an accident, the metallic implant plays a crucial role in replacing the natural bone. Because the structural implant has a more significant porosity, the AM method encourages cell proliferation and tissue ingrowth. This technology is particularly useful in biomedical applications since it incorporates computer aided design (CAD) data into a personalized physical implant that is tailored to the anatomy of the individual patient. This approach is especially crucial when employing 3D printing to create orthoses, such as a fracture-specific cast (33,34).

7.4.3.2 Dentistry

Dental implants require very precise form and size to be accumulated at the maxillofacial/endodontics location, making traditional production of dental implants a difficult process. As a result, AM drew a lot of interest from researchers in order to produce very accurate customer-specific dental implants. AM plays an important role in tailoring the biocompatible and osteoinductive model in oral surgery. Stereolithography is one of the well-known processes for developing commercial resin products. This process is also beneficial in stimulating bone regeneration. PBF is also a viable method for fabricating dental implants with precise form and size (35,36).

7.4.4 Pharmaceuticals

In the pharmaceutical industry, AM plays a vital role in medication and delivery applications. It boosts medication release through several channels, allowing the patient to recover more quickly. Fused deposition modeling (FDM) is a low-cost, high-speed method that is commonly employed in the production of tablets. Using a hydroalcoholic gel–based paste, the FDM can print the tablet compartment in three pieces. The three compartment-based pills can be used for a variety of medicinal purposes. The traditional approach has a significant barrier in producing three compartments in a single capsule. Only AM can provide a wide range of different dosages in a single dose. The main aim in this regard is to manufacture patch and microneedle arrays that help the continuous supply of the drug to the upper layer of the epidermis. In the pharmaceutical industry, AM has a number of benefits, including customization and personalization, high productivity, high resolution, high precision, high reliability, cheap production costs, the capacity to adjust the size of droplets, high dose strength, multi dosages, and so on (37–39).

7.4.5 Surgical tools

AM is a very effective tool that can convert the CAD file into a usable tool that can serve the ongoing needs of medical institutions and hospital operating rooms. AM allowed the designing and development of patient-specific surgical tools and instruments. However, the complex anatomical procedure can be more controlled and simplified using customized tools (40).

7.4.6 Operative planning

It is one of the most well-known applications of AM in operative planning. Stereolithography (SLA) technology may provide a patient's anatomy picture that might help surgeons improve their eyesight while repairing any damaged tissues. Finally, this strategy can help the surgeon design a successful surgery (41).

7.4.7 Implant tissue interface

Stress shielding is the most common cause of implant failure over time owing to a mismatch of the young's modulus between bones and the implanted parts. Due to the constant movement of the implant tissue surface, the metallic implants produced cell death. As a result, it is critical to evaluate the surface properties of the implant that help in better integration and attachment to surrounding tissues. The roughness of the surface aids in cell attachment to the bone. Weißmann et al. created a 3D printed titanium bone utilizing an Electron beam melting (EBM) printed with a high roughness value in their investigation (42). The in vitro results demonstrated that the implant surface had good osteoblast proliferation and cell viability.

7.4.8 Personalized protective equipment during COVID 19

The entire globe is presently dealing with the CORONA virus (a second mutation of the SARS-2 virus), which is a dreadful infectious disease. Many international and national health agencies or organizations advised using tailored protective equipment (PPE) for respiratory, eye, and hand protection to shield interactions with COVID 19 patients. Because health care staff had to care for sick patients during the COVID 19 crisis, health care facilities were more risky. The entire supply chain has been interrupted as a result of the shutdown owing to COVID-19 virus, including medical and health care items such as manifold connected helmets, face masks, oxygen PPE masks, face shields, and oxyframe attached PPE, as depicted in Figure 7.3. With the assistance of scientists and researchers, 3D printing technology was actively involved in the fabrication of various PPE components like face shields, ventilator valves, face masks, and hand gloves at that time. Some of the PPE is covered below.

Role of additive manufacturing   125

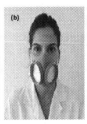

Figure 7.3 Additively manufactured PPE components: (a) manifold attached helmet, (b) P2 half mask, (c) oxygen PPE mask, (d) face shield, and (e) oxyframe attached PPE (28).

Table 7.3 Advantages of additive manufacturing over conventional manufacturing (43)

Area of application	Advantages
Customized items (33)	Mass customization at a minimal cost Great customer satisfaction Customized replacement of parts on site
Cost-effectiveness (34)	Small production at low cost Eliminate the tooling cost Low labor cost Eliminate costly warehousing Reduce product development cost
Higher productivity (35)	No delay time No tooling required No rework Relatively inexpensive for production of a small number of components
Complex items (36)	Highly complex parts can be produced at a very low cost No design constraints
Quick repair (37)	Very low repair time Modified the repaired items with a new design
Onsite and on-demand manufacturing (38)	Eliminate the storage and transportation cost Shorten supply chain No large inventory is needed
Higher quality (39)	The higher resolution of the parts Implants and prosthetics are produced with high quality
Flexibility (40)	Higher flexibility to change material

7.5 ADVANTAGES OF ADDITIVE MANUFACTURING OVER CONVENTIONAL MANUFACTURING IN THE BIOMEDICAL FIELD

AM is a classic process that has the potential to become the most remarkable technique in the coming years. AM is a really unique approach that offers new prospects for the biomedical sectors due to its several benefits over traditional production techniques, including customization, cost, productivity, flexibility, and quality. Table 7.3 highlights the advantages of the AM approach over the older techniques in the next section.

7.6 LIMITATIONS OF ADDITIVE MANUFACTURING IN THE BIOMEDICAL DOMAIN

Over the last few years, AM techniques have received a lot of attention. Despite its benefits, AM has a number of scientific, technical, and economic restrictions. The following are in-depth discussions of the aforementioned issues:

7.6.1 Cost-effective only for low-volume production

AM is the most effective technique if the part produced by this technique is being used for a particular application; otherwise, the process would be wastage of material (44).

7.6.2 Limited material option

Material selection for biomedical purposes is exceptionally complex in AM as every material is not compatible with every AM technology (45).

7.6.3 Poor mechanical properties

Some AM techniques are not suitable for biomedical applications due to the insufficient mechanical strength of the fabricated parts after the printing process (46).

7.6.4 Low-dimensional accuracy

It is crucial for medical components to appropriately fit into the human body, which is not always achievable with additive-made items. The 3D models created by AM lack dimensional precision and have poor surface polish (47).

7.7 MECHANICAL PROPERTIES OF BIOMEDICAL PARTS

Mechanical properties, such as tensile and compressive strength, fatigue strength, fracture toughness, surface roughness, and residual micro strains and stresses, all have a significant impact on the life of a metallic implant (48). Mechanical properties can vary depending on various parameters such as material, fabrication procedures, laser power, temperature, ambient conditions, and scan or deposition approach, among others (49,50).

In this section, mechanical qualities, such as tensile and compressive strength, as well as fracture toughness of 3D-printed medical components, are discussed:

7.7.1 Tensile and compressive strength

Material properties, manufacturing techniques, and processing factors all have an impact on the tensile and compressive strength of metallic components. For biomedical components, the tensile test is often conducted at room temperature. For example, the ultimate tensile strength of laser metal deposition (LMD)-based Ti6Al4V is between 920 and 1150 MPa (51).

7.7.2 Fracture toughness

The capacity of a flaw-containing material to withstand an applied load is referred to as fracture toughness (KiC). The fracture toughness value is affected by the processing conditions, microstructure, and roughness of the component. The fracture toughness value of LMD Ti6Al4V alloys, for example, is substantially higher than that of wrought Ti6Al4V alloys, but it is quite near to that of the EBM method. On the additively created Ti6Al4V specimen, dimpled and smooth portions were detected. The dimpled section has more plastic deformation before failure, whereas the flat area has a brittle failure (52).

7.8 FUTURE ASPECTS

Bioprinting offers the ability to produce tissue-like 3D models of physiological systems for drug screening and disease modeling. Bioprinting may be used to manufacture organoids, which can be directly transplanted into patients. Waiting lists or thorough histocompatibility testing, as well as the patient's own cells, can be employed in the bioprinting process. Before attempting to print a whole organ, the bioprinting process can overcome obstacles such as micro vascularization and bioink lifetime. 3D printing is a powerful technology for creating scaffolds with specified geometries and tailored porous structures. Variable construction parameters like power and scanning speed can be used to accomplish the aforementioned qualities. 3D printing remains a significant problem in terms of optimizing higher quality components that may be employed in biomedical applications. Because porosity is important in scaffolding and metallic implants, a highly porous construction was required. Loose particles that are caught inside the pores are extremely difficult to dislodge. After sintering, the loose particles may

128 Additive Manufacturing

minimize the size of the holes, increasing the strength of the scaffold or metallic implant and reducing cell infiltration inside the pores.

7.9 CONCLUSION

AM is a rapidly growing technique, particularly in the biomedical arena. This approach has a wide range of applications, including medicines, surgical planning models, and high-accuracy bespoke items. Furthermore, AM is linked to recent breakthroughs such as surgical models, tailored prostheses, patient-specific surgical instruments, tissue engineering and tissue production, orthopedic implants, dentistry, and medicines. A comprehensive study of the benefits, drawbacks, and applications in many biomedical sectors has been conducted in this study.

REFERENCES

1. Kumar R, Kumar M, Chohan JS. The role of additive manufacturing for biomedical applications: A critical review. *J Manuf Process [Internet]*. 2021;64(September 2020):828–50. Available from: https://doi.org/10.1016/j.jmapro.2021.02.022.
2. Norman J, Madurawe RD, Moore CMV, Khan MA, Khairuzzaman A. A new chapter in pharmaceutical manufacturing: 3D-printed drug products. *Adv Drug Deliv Rev.* 2017;108:39–50.
3. Coelho G, Chaves TMF, Goes AF, Del Massa EC, Moraes O, Yoshida M. Multimaterial 3D printing preoperative planning for frontoethmoidal meningoencephalocele surgery. *Child's Nerv Syst.* 2018;34(4):749–56.
4. Yang C, Wang X, Ma B, Zhu H, Huan Z, Ma N, et al. 3D-printed bioactive Ca3SiO5 bone cement scaffolds with nano surface structure for bone regeneration. *ACS Appl Mater Interfaces.* 2017;9(7):5757–67.
5. Rosenzweig DH, Carelli E, Steffen T, Jarzem P, Haglund L. 3D-printed ABS and PLA scaffolds for cartilage and nucleus pulposustissue regeneration. *Int J Mol Sci.* 2015;16(7):15118–35.
6. Nives cubo, marta gracia jua. 3D bioprinting of functional human skin: Production and invivo analysis. *Biofabrication.* 2016;9(1):1–12.
7. Baumers M, Dickens P, Tuck C, Hague R. The cost of additive manufacturing: Machine productivity, economies of scale and technology-push. *Technol Forecast Soc Change [Internet]*. 2016;102:193–201. Available from: http://dx.doi.org/10.1016/j.techfore.2015.02.015.
8. Liu S, Shin YC. Additive manufacturing of Ti6Al4V alloy: A review. *Mater Des [Internet]*. 2019;164:107552. Available from: https://doi.org/10.1016/j.matdes.2018.107552.
9. Salmi M. Additive manufacturing processes in medical applications. *Materials (Basel).* 2021;14(1):1–16.
10. Jaber H. Selective laser melting of Ti alloys and hydroxyapatite for tissue engineering: Progress and challenges. *Mater Res Exp.* 2019;6(8):082003.

Role of additive manufacturing 129

11. Singh S, Ramakrishna S, Singh R. Material issues in additive manufacturing: A review. *J Manuf Process [Internet]*. 2017;25:185–200. Available from: http://dx.doi.org/10.1016/j.jmapro.2016.11.006.

12. Jiang J, Fu YF. A short survey of sustainable material extrusion additive manufacturing. *Aust J Mech Eng [Internet]*. 2020;00(00):1–10. Available from: https://doi.org/10.1080/14484846.2020.1825045.

13. Bose S, Ke D, Sahasrabudhe H, Bandyopadhyay A. Progress in materials science additive manufacturing of biomaterials. *Prog Mater Sci [Internet]*. 2018;93:45–111. Available from: https://doi.org/10.1016/j.pmatsci.2017.08.003.

14. Gunatillake PA, Adhikari R, Gadegaard N. Biodegradable synthetic polymers for tissue engineering. *Eur Cells Mater*. 2003;5:1–16.

15. Hutmacher DW, Schantz T, Zein I, Ng KW, Teoh SH, Tan KC. Mechanical properties and cell cultural response of polycaprolactone scaffolds designed and fabricated via fused deposition modeling. *J Biomed Mater Res*. 2001;55(2):203–16.

16. Williams JM, Adewunmi A, Schek RM, Flanagan CL, Krebsbach PH, Feinberg SE, et al. Bone tissue engineering using polycaprolactone scaffolds fabricated via selective laser sintering. *Biomaterials*. 2005;26(23):4817–27.

17. Lee M, Dunn JCY, Wu BM. Scaffold fabrication by indirect three-dimensional printing. *Biomaterials*. 2005;26(20):4281–9.

18. lodhi MJK, Deen KM, Wacker Greenlee MC, Haider W. Additive Manufactured 316L stainless steel with improved corrosion resistance and biological response for biomedical applications. *Addit Manuf*. 2019;27:8–19.

19. Ponader S, Vairaktaris E, Heinl P, Wilmowsky CV, Rottmair A, Körner C, et al. Effects of topographical surface modifications of electron beam melted Ti-6Al-4V titanium on human fetal osteoblasts. *J Biomed Mater Res – Part A*. 2008;84(4):1111–9.

20. Gaytan SM, Murr LE, Martinez E, Martinez JL, MacHado BI, Ramirez DA, et al. Comparison of microstructures and mechanical properties for solid and mesh cobalt-base alloy prototypes fabricated by electron beam melting. *Metall Mater Trans A Phys Metall Mater Sci*. 2010;41(12):3216–27.

21. Kumar A, Pandey PM. Effect of ultrasonic assisted sintering on mechanical properties and degradation behaviour of Mg15Nb3Zn1Ca biomaterial. *J Magnes Alloy [Internet]*. 2021;9(6):1989–2008. Available from: https://doi.org/10.1016/j.jma.2020.11.006.

22. Canillas M, Pena P, De Aza AH, Rodríguez MA. Calcium phosphates for biomedical applications. *Bol la Soc Esp Ceram y Vidr [Internet]*. 2017;56(3):91–112. Available from: http://dx.doi.org/10.1016/j.bsecv.2017.05.001.

23. Szcześ A, Hołysz L, Chibowski E. Synthesis of hydroxyapatite for biomedical applications. *Adv Colloid Interface Sci*. 2017;249(April):321–30.

24. Vichery C, Nedelec JM. Bioactive glass nanoparticles: From synthesis to materials design for biomedical applications. *Materials (Basel)*. 2016;9(4).

25. Xynos ID, Edgar AJ, Buttery LDK, Hench LL, Polak JM. Gene-expression profiling of human osteoblasts following treatment with the ionic products of Bioglass® 45S5 dissolution. *J Biomed Mater Res*. 2001;55(2):151–7.

26. Eosoly S, Brabazon D, Lohfeld S, Looney L. Selective laser sintering of hydroxyapatite/poly-ε-caprolactone scaffolds. *Acta Biomater*. 2010;6(7):2511–7.

27. Korpela J, Kokkari A, Korhonen H, Malin M, Narhi T, Seppalea J. Biodegradable and bioactive porous scaffold structures prepared using fused deposition modeling. *J Biomed Mater Res – Part B Appl Biomater.* 2013;101(4):610–9.
28. Boyan BD, Hummert TW, Dean DD, Schwartz Z. Role of material surfaces in regulating bone and cartilage cell response. *Biomaterials.* 1996;17(2):137–46.
29. Leong KF, Cheah CM, Chua CK. Solid freeform fabrication of three-dimensional scaffolds for engineering replacement tissues and organs. *Biomaterials.* 2003;24(13):2363–78.
30. Jafferson JM, Pattanashetti S. Use of 3D printing in production of personal protective equipment (PPE) – A review. *Mater Today Proc [Internet].* 2021;46:1247–60. Available from: https://doi.org/10.1016/j.matpr.2021.02.072.
31. Moreno Madrid AP, Vrech SM, Sanchez MA, Rodriguez AP. Advances in additive manufacturing for bone tissue engineering scaffolds. *Mater Sci Eng C [Internet].* 2019;100(March 2018):631–44. Available from: https://doi.org/10.1016/j.msec.2019.03.037.
32. Albrektsson T, Johansson C. Osteoinduction, osteoconduction and osseointegration. *Eur Spine J.* 2001;10:S96–101.
33. Cartiaux O, Paul L, Francq BG, Banse X, Docquier PL. Improved accuracy with 3D planning and patient-specific instruments during simulated pelvic bone tumor surgery. *Ann Biomed Eng.* 2014;42(1):205–13.
34. Pati F, Song TH, Rijal G, Jang J, Kim SW, Cho DW. Ornamenting 3D printed scaffolds with cell-laid extracellular matrix for bone tissue regeneration. *Biomaterials [Internet].* 2015;37:230–41. Available from: http://dx.doi.org/10.1016/j.biomaterials.2014.10.012.
35. Oberoi G, Nitsch S, Edelmayer M, Janjic K, Müller AS, Agis H. 3D printing-Encompassing the facets of dentistry. *Front Bioeng Biotechnol.* 2018;6(NOV):1–13.
36. Hao W, Liu Y, Zhou H, Chen H, Fang D. Preparation and characterization of 3D printed continuous carbon fiber reinforced thermosetting composites. *Polym Test [Internet].* 2018;65(September 2017):29–34. Available from: http://dx.doi.org/10.1016/j.polymertesting.2017.11.004.
37. Pardeike J, Strohmeier DM, Schrödl N, Voura C, Gruber M, Khinast JG, et al. Nanosuspensions as advanced printing ink for accurate dosing of poorly soluble drugs in personalized medicines. *Int J Pharm [Internet].* 2011;420(1):93–100. Available from: http://dx.doi.org/10.1016/j.ijpharm.2011.08.033.
38. Scoutaris N, Alexander MR, Gellert PR, Roberts CJ. Inkjet printing as a novel medicine formulation technique. *J Control Release [Internet].* 2011;156(2):179–85. Available from: http://dx.doi.org/10.1016/j.jconrel.2011.07.033.
39. Goyanes A, Buanz ABM, Hatton GB, Gaisford S, Basit AW. 3D printing of modified-release aminosalicylate (4-ASA and 5-ASA) tablets. *Eur J Pharm Biopharm [Internet].* 2015;89:157–62. Available from: http://dx.doi.org/10.1016/j.ejpb.2014.12.003.
40. Derakhshanfar S, Mbeleck R, Xu K, Zhang X, Zhong W, Xing M. 3D bioprinting for biomedical devices and tissue engineering: A review of recent trends and advances. *Bioact Mater [Internet].* 2018;3(2):144–56. Available from: https://doi.org/10.1016/j.bioactmat.2017.11.008.

Role of additive manufacturing **131**

41. Wake N, Alexander AE, Christensen AM, Liacouras PC, Schickel M, Pietila T, et al. Creating patient-specific anatomical models for 3D printing and AR/VR: A supplement for the 2018 Radiological Society of North America (RSNA) hands-on course. *3D Print Med.* 2019;5(1).

42. Weißmann V, Drescher P, Seitz H, Hansmann H, Bader R, Seyfarth A, et al. Effects of build orientation on surface morphology and bone cell activity of additively manufactured Ti6Al4V specimens. *Materials (Basel).* 2018;11(6).

43. Pereira T, Kennedy JV, Potgieter J. A comparison of traditional manufacturing vs additive manufacturing, the best method for the job. *Procedia Manuf [Internet].* 2019;30:11–8. Available from: https://doi.org/10.1016/j.promfg.2019.02.003.

44. Scott J, Gupta N, Weber C, Newsome S. Additive manufacturing: Status and opportunities. *Sci Technol Policy Institute, Washington.* 2012.

45. Attaran M. The rise of 3-D printing: The advantages of additive manufacturing over traditional manufacturing. *Bus Horiz [Internet].* 2017;60(5): 677–88. Available from: http://dx.doi.org/10.1016/j.bushor.2017.05.011.

46. Hermawan H, Ramdan D, P. Djuansjah JR. Metals for biomedical applications. *Biomed Eng – From Theory to Appl.* 2011.

47. Ngo TD, Kashani A, Imbalzano G, Nguyen KTQ, Hui D. Additive manufacturing (3D printing): A review of materials, methods, applications and challenges. *Compos Part B Eng [Internet].* 2018;143(December 2017):172–96. Available from: https://doi.org/10.1016/j.compositesb.2018.02.012.

48. Mitsuo N. Mechanical properties of biomedical titanium alloys. *Mater Sci Eng A [Internet].* 1998;243(1–2):231–6. Available from: http://www.sciencedirect.com/science/article/pii/S092150939700806X.

49. Kumar A, Pandey PM. Study of the influence of microwave sintering parameters on the mechanical behaviour of magnesium-based metal matrix composite. *Proc Inst Mech Eng Part C J Mech Eng Sci.* 2021;235(13):2416–25.

50. Kumar A, Pandey PM. Comparison of processing routes for efficacious fabrication of Mg3Zn1Ca15Nb biomaterial. *Mater Manuf Process [Internet].* 2021;36(12):1365–76. Available from: https://doi.org/10.1080/10426914.20 21.1914846.

51. Zhong C, Liu J, Zhao T, Schopphoven T, Fu J, Gasser A, et al. Laser metal deposition of Ti6Al4V – A brief review. *Appl Sci.* 2020;10(3):1–12.

52. Azarniya A, Colera XG, Mirzaali MJ, Sovizi S, Bartolomeu F, St Weglowski MK, et al. Additive manufacturing of Ti–6Al–4V parts through laser metal deposition (LMD): Process, microstructure, and mechanical properties. *J Alloys Compd [Internet].* 2019;804:163–91. Available from: https://doi.org/10.1016/j.jallcom.2019.04.255.

Chapter 8

Wire arc additive manufacturing of titanium alloys

A review on properties, challenges, and applications

Sujeet Kumar
National Institute of Technology (Patna)

Vimal K.E.K.
National Institute of Technology (Tiruchirappalli)

CONTENTS

8.1	Introduction	134
8.2	Wire arc additive manufacturing techniques	135
	8.2.1 Gas tungsten arc welding process	136
	8.2.2 Gas metal arc welding process	137
8.3	Wire arc additive manufacturing of titanium alloys	138
	8.3.1 Titanium alloys fabricated using gas tungsten arc welding–based wire arc additive manufacturing	139
	8.3.1.1 Microstructure analysis	139
	8.3.1.2 Tensile properties analysis	139
	8.3.2 Titanium alloys fabricated using gas tungsten arc welding–based wire arc additive manufacturing	140
	8.3.2.1 Microstructure analysis	140
	8.3.2.2 Tensile properties analysis	140
	8.3.3 Titanium alloys fabricated using cold metal transfer–based wire arc additive manufacturing	140
	8.3.3.1 Microstructure analysis	141
	8.3.3.2 Tensile properties analysis	141
	8.3.4 Challenges during wire arc additive manufacturing process for producing Titanium-alloyed parts	141
	8.3.5 Common defects in titanium alloys produced by wire arc additive manufacturing techniques	142
	8.3.6 Strategies for quality improvement	142
8.4	Conclusion	142
References		143

DOI: 10.1201/9781003258391-8

8.1 INTRODUCTION

Additive manufacturing (AM), which is a good name for technologies that manufacture 3D products by layering material on top of each other, is quickly gaining traction in the manufacturing sector [1–5]. AM is employed in industrial manufacturing because of its high deposition rate, low material waste, low equipment cost, and environmental friendliness [5–10]. Furthermore, AM technologies can be classified as either a powder-feed approach or a wire-feed method in terms of how the added material is delivered [11,12].

Wire arc additive manufacturing (WAAM), as one of the AM methods, is a suitable option for producing large metal parts from costly materials [13,14]. The heat source in a WAAM method is commonly derived from gas tungsten arc welding (GTAW) or gas metal arc welding (GMAW) [15–17]. The feedstock materials are generally commercial welding wires, and the motion mechanism is usually an articulated production robot to have a texture deposition path to manufacture full-density metal pieces [18,19]. To generate an arc, a nonconsumable tungsten electrode is required for GTAW, whereas GMAW employs the deposition wire as an electrode [20,21]. The GMAW process has a simple heat source to create the melt pool because of coaxial wire feeding through the center of the welding torch. In the GTAW process, heat wire is fed separately to the weld pool with some angle with a welding torch [21,22]. The deposition rate of GMAW-based WAAM is two to three times that of GTAW approaches. The GMAW-based WAAM is less stable and generates more welding smoke or spatter when an electric current is applied directly to the material [23,24]. The WAAM approach has been proven to be capable of producing high-quality Ti6Al4V components with outstanding mechanical properties that are comparable to those generated using traditional methods. The cold metal transfer (CMT) process is a variation of the GMAW method developed by Fronius. It uses a controlled dip transfer mode for metal transfer. WAAM process through CMT results from a high metal deposition rate with minimum heat input [25]. Due to low heat input, CMT offers minimum residual stress, low distortion, and minimizes the problem related to heat collapse during the process [14,26].

Metal alloys like stainless steel, nickel-based alloys, aluminum alloys, and other structural metal alloys have all been employed in the WAAM process. Titanium alloy offers outstanding overall qualities, such as high temperature resistance, light in weight, high toughness, large ultimate strength, high fatigue strength, higher propagation of crack resistance, and corrosion resistance among the structural alloys [27]. Among all metals, titanium alloy has attracted a lot of attention due to its vast range of uses in aircraft and aerospace, defense and weaponry, maritime and shipyard, automobile, and bioengineering [28–30]. Li et al. employed the GTAW WAAM approach to make titanium alloy and discovered that surface peening technology is effective for microstructure modification after the WAAM process because the size of the grain reduces [31]. In this paper, the WAAM techniques (GTAW, GMAW, and CMT) are analyzed for titanium alloys. Also, the microstructural and mechanical characteristics of titanium alloys produced by WAAM are reviewed.

8.2 WIRE ARC ADDITIVE MANUFACTURING TECHNIQUES

WAAM is a metallic AM process that feeds the material with metal wires and employs an electric arc as a heat source. It has many advantages like greater material deposition rate, lower manufacturing lead time, and larger material utilization [31–33]. It utilizes basic traditional welding methods (GTAW, GMAW) for the heat source and wire feeding. The flow chart of the WAAM technique is given in Figure 8.1. Schematic diagram that represents various components used in the WAAM processes is shown in Figure 8.2.

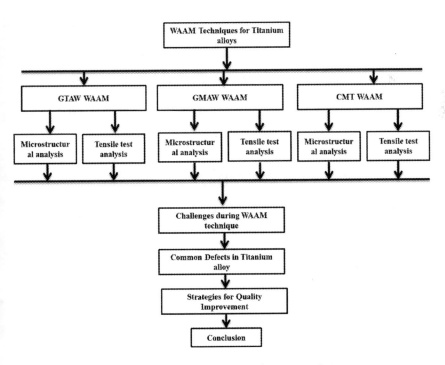

Figure 8.1 Flow chart of the wire arc additive manufacturing technique.

136 Additive Manufacturing

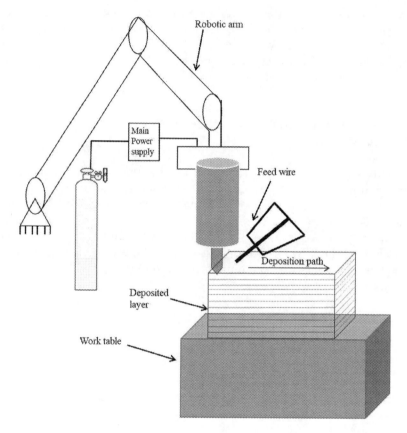

Figure 8.2 Schematic diagram of the wire arc additive manufacturing process.

8.2.1 Gas tungsten arc welding process

As demonstrated in Figure 8.3, the gas tungsten arc welding (GTAW) technique uses a nonconsumable tungsten electrode for welding. To speed up the deposition process, filler wire might be employed. GTAW is suitable to weld titanium alloy, aluminum, nickel alloy, and stainless steel family. To protect the weld bead from external contamination, shielding gas (Ar, Co_2, etc.) is used [34]. Metal deposition and orientation affect the material transfer and the quality of the weld. Front feeding is used for titanium alloys [24]. GTAW twin-wire WAAM technologies are used to increase the weld deposition and manufacture the functionally graded and intermetallic material [35].

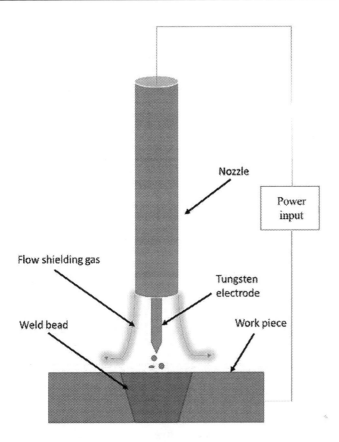

Figure 8.3 Gas tungsten arc welding process setup.

8.2.2 Gas metal arc welding process

A consumable electrode wire was used in the gas metal arc welding (GMAW) process to feed in the weld pool, as shown in Figure 8.4. Shielding gas (Ar, Co_2) is provided to protect the weld from being contaminated by the environment [12]. The shielding gas used and the deposition rate in different WAAM techniques are given in Table 8.1. This process is preferred for its faster welding speed and high deposition rate. GMAW can transfer metals by globular, short-circuiting, and spray transfer modes [34]. GMAW-based WAAM process can achieve a metal deposition rate of up to 8 kg/h [36,37].

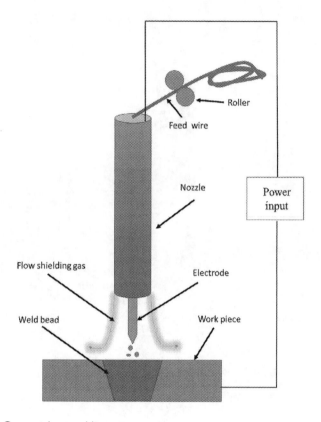

Figure 8.4 Gas metal arc welding process setup.

Table 8.1 Shielding gas used and deposition rate in different wire arc additive manufacturing techniques [21]

Method	Shielding gas used	Deposition rate (kg/h)
Gas tungsten arc welding (GTAW) wire arc additive manufacturing (WAAM) process	He (70%) and Ar (30%)	1.0
GMAW WAAM process	Pure Ar He (50%) and Ar (50%)	2.5
Cold metal transfer (CMT)	Pure He	2.5

8.3 WIRE ARC ADDITIVE MANUFACTURING OF TITANIUM ALLOYS

Metal alloys like stainless steel, nickel-based alloys, aluminum alloys, and other structural metal alloys have all been employed in the WAAM process. Titanium alloy offers outstanding overall qualities, such as high temperature

Wire arc additive manufacturing of titanium alloys 139

Table 8.2 Chemical composition of titanium alloy and filler wire [28]

Materials	Al	V	Fe	C	N	H	O	Ti
Titanium alloys	6.05	4.02	0.14	0.02	0.021	0.006	0.12	Balance
Filler wire	6.01	3.93	0.18	0.08	0.01	0.002	0.09	Balance

resistance, light in weight, high toughness, large ultimate strength, high crack propagation resistance, and resistance to corrosion, among the structural alloys [27]. The chemical composition of titanium alloy and filler metal wire is given in Table 8.2. Among all material properties, titanium alloy has attracted a lot of attention due to its vast range of uses in aircraft and aerospace, defense and weaponry, maritime and shipyard, automobile, and bioengineering [38,39]. Li et al. employed the GTAW WAAM approach to make titanium alloy and discovered that surface peening technology is effective for microstructure modification after the WAAM process because the size of the grain reduces [31].

8.3.1 Titanium alloys fabricated using gas tungsten arc welding–based wire arc additive manufacturing

The GTAW welding process is used for the WAAM of titanium alloys because of less cost, easy availability, and better microstructural and mechanical properties.

8.3.1.1 Microstructure analysis

The microstructure of WAAM manufactured titanium alloy depends on the cooling rate, solidification time, and process temperature [B. Wu]. GTAW process–based WAAM manufactured titanium alloys show columnar prior β grains and Widmanstatten α/β grains due to less solidification time. Baufeld et al. reported α phase lamella basket weave structures of titanium alloys [12]. Wang et al. analyzed the microstructures and found Widmanstatten α and coarsened structure lamella (α) [41]. WAAM manufactured titanium alloys post-heat treatment show great improvement in the microstructures. Heat treatment at 600°C for 4 hours of manufactured titanium alloys gives a lamellar structure. Heat treatment of manufactured titanium alloys for 2 hours at 834°C shows a lamellar structure [12,35,40–42].

8.3.1.2 Tensile properties analysis

Tensile properties (yield strength (YS) and ultimate tensile strength (UTS)) of WAAM manufactured titanium alloys are degraded due to minimum spacing of α lamellae boundary and high interpass temperature (200°C) [42]. Baufeld et al. reported UTS of 929 MPa for GTAW WAAM manufactured

140 Additive Manufacturing

titanium alloys [12]. Wang et al. reported YS and UTS of 803 and 918 MPa, respectively. After heat treatment (600°C for 4 hours) of WAAM, the manufactured component's tensile strength is increased to 972 MPa [41]. Brandl et al. indicated the YS of 861 MPa and the UTS of 937 MPa [12].

8.3.2 Titanium alloys fabricated using gas tungsten arc welding–based wire arc additive manufacturing

The GMAW process uses high voltage among the substrate and consumable electrode to produce the arc. The molten consumable wire makes the layering on the metal [43]. Panchenko et al. established an efficient and controlled mode of metal transfer (short-circuiting) by changing the voltage and arc current which decreased the heat input by 16% [44].

8.3.2.1 Microstructure analysis

The microstructure of GMAW-based WAAM manufactured titanium alloys indicates epitaxial columnar β grains grown on the deposited wall through the substrate. The top portion of the deposited layer shows a fine α+β Widmanstatten structure and the inside layer shows coarse Widmanstatten α colonies due to fast cooling and high temperature [41]. Minor lamellar α+β and acicular α' martensitic grains are found during GMAW-based WAAM process. The bottom edge of the wall contains a coarse grain structure because of the lower temperature gradient. The middle portion of the deposited layer contains columnar grains and the top portions have a fine grain structure [44]. Heat treatment after layering can improve microstructural and mechanical characteristics of manufactured components. Post-heat treatment of titanium alloys, the manufactured components (900°C for 4 hours and 1200°C for 2 hours) changed the microstructure to lamellar α+β mixture [22,45].

8.3.2.2 Tensile properties analysis

Greater strength and slightly less ductility are found in a horizontal oriented tensile test as compared to an extruded bar of titanium alloys [41]. Paskual et al. reported the YS and UTS of 820 and 960 MPa, respectively. Post-heat treatment (at 900°C for 4 hours), the UTS is reduced to 841 MPa [45]. The strain stress curve for GMAW-based WAAM manufactured titanium alloy.

8.3.3 Titanium alloys fabricated using cold metal transfer–based wire arc additive manufacturing

CMT is a modification of the GMAW process developed by Fronius. It uses a controlled dip transfer mode for metal transfer. WAAM process through CMT results in a high metal deposition rate with minimum heat input [26].

Wire arc additive manufacturing of titanium alloys 141

8.3.3.1 Microstructure analysis

Microstructures of titanium alloys tend to form coarse-columnar β grain during CMT manufacturing because of the continuous solidification process and temperature gradient [46, 47]. The top layer of CMT manufactured titanium alloys shows the martensitic structure (α) and a fine lamella of α phase. The middle portion α phase started increasing and forming colonies. The bottom portion shows lamellae structure grown from the α phase due to the effect of cooling rate [48]. At a low cooling rate, α grain boundary formed before β grains. A higher cooling rate α is formed like a basket wave structure inside the β grain. Very high cooling rate (525°C/s) formed a martensitic structure (α) [49]. Fusion boundary indicates coarse columnar produced from the β phase. After cooling the β phase, a combined microstructure of α-lamella and α+β consisting of Widmanstatten colony is observed [50]. Coarse-columnar grain grown from the β phase can be rearranged with rolling operation after each layer deposition during the CMT-based WAAM process. Rolling operation refines the coarse-columnar grain which enhances the weld's microstructural and mechanical qualities [51–55].

8.3.3.2 Tensile properties analysis

The YS and UTS of CMT manufactured titanium alloys show a value of 958 and 1046 MPa, respectively. Post-heat treatment (710°C, 850°C, and 920°C) of manufactured components increased the yield strength but slightly decreased the UTS due to grain rearrangement [48]. Zhang et al. reported that the UTS drops from 984.5 to 899.2 MPa when the deposition rate is increased from 1.63 to 2.23 kg/h [50]. Martina et al. found that fabricated titanium alloy's yield strength and UTS were 807 and 903 MPa, respectively. Interpass cold rolling increased the yield strength and UTS to 916 and 1022 MPa, respectively [33]. The addition of niobium (Nb) powder improved the UTS of the CMT manufactured sample. The stress-strain curve shows the as-built sample and Nb added sample UTS and elongation [55].

8.3.4 Challenges during wire arc additive manufacturing process for producing Titanium-alloyed parts

WAAM has several factors to consider for manufacturing the titanium alloys parts such as reduction of heat accumulation during layering, parameter setup, good programming plan, stable weld pool dynamics, avoiding gas contamination, and thermal deformation [17,38]. WAAM manufactured components go through high heat input with heat accumulation in the layered. This can change the microstructure undesirably and drastically change the mechanical properties like tensile strength due to residual stress generation [52,53].

8.3.5 Common defects in titanium alloys produced by wire arc additive manufacturing techniques

Several factors, such as the arc welding method, process variables, air velocity, interpass temperatures, wire condition, and alloy type, create porosity [54–57]. Plate distortion is among the most typical flaws in the WAAM process (angular distortion, bending distortion, and rotational distortion), involving longitudinal and transverse shrinkage. The residual stress causes delamination of manufactured layered, fatigue strength degradation, and fabricated component fatigue resistance. Therefore, minimization of deformation and residual stress is very important for better mechanical and microstructural properties of the manufactured parts [58,59].

8.3.6 Strategies for quality improvement

The microstructure of WAAM-produced components can be enhanced, and residual stress developed by thermal (after heat treatment) and mechanical diagnosis can be reduced [52,53]. The problem of oxidation of titanium alloys in the open air can be prevented by trailing gas shielding. Preheating can control the interpass temperature during the WAAM manufacturing technique [60]. Mechanical strength and microstructural performance can be enhanced by iso-static processing and thermomechanical processing apart from a post-heat treatment [33]. The residual stress produced due to the WAAM process on a different part of the manufactured components can be eliminated by the post-heat-treatment process. Residual stress is compressive in nature and can help to increase the fatigue life, resist the crack formation, and increase corrosion resistance [61,62]. The mechanical peening technique makes the compressive stresses. Ultrasonic impact treatments (UIT) technique releases the tensile residual stress by applying compressive strength and can reduce the developed residual stress and enhance the mechanical properties [63]. A compressed CO_2-based forced interpass cooling is used to produced titanium alloys. Inter-pass cooling reduced the oxidation, improved the hardness value, refined the microstructure, and enhanced the mechanical strength of fabricated titanium alloy due to minimizing the dwell time among the deposited layers [64–70].

8.4 CONCLUSION

In this review paper, WAAM techniques (GTAW, GMAW, and CMT) for producing titanium alloys are discussed. Microstructural studies and tensile strength analysis are done based on many researchers' papers.

GTAW process–based WAAM manufactured titanium alloys' microstructure revealed columnar prior β grains and Widmanstatten α/β grains because of fast cooling rate and UTS found in the range of 918–929 MPa. Heat treatment (600°C for 4 hours) increased the UTS to 972 MPa. GTAW process–based WAAM manufactured titanium alloys' microstructure shows coarse Widmanstatten α colonies fast cooling and high temperature. Tensile strength was found at 960 MPa which is higher than the GTAW-based WAAM process. CMT-based WAAM process revealed martensitic structure (α) and lamellae microstructure. The tensile strength of CMT-based WAAM was 903 MPa and after Inter pass cold rolling increased it to 1022 MPa. The WAAM manufacturing process is challenging in many aspects like parameter setup, heat accumulation during layering, and gas contamination. Common defects in the produced titanium alloys are porosity, plate distortion (angular distortion, bending distortion, and rotational distortion), longitudinal and transverse shrinkage, and residual stress. Thermal (after heat treatment) and mechanical treatment (UIT) can improve the quality of WAAM-produced components.

REFERENCES

1. Wong, K. V., & Hernandez, A. (2012). A review of additive manufacturing. *International Scholarly Research Notices*.
2. Frazier, W. E. (2014). Metal additive manufacturing: A review. *Journal of Materials Engineering and Performance*, 23(6), 1917–1928.
3. Herzog, D., Seyda, V., Wycisk, E., & Emmelmann, C. (2016). Additive manufacturing of metals. *ActaMaterialia*, 117, 371–392.
4. Gibson, I., Rosen, D. W., Stucker, B., Khorasani, M., Rosen, D., Stucker, B., & Khorasani, M. (2021). *Additive Manufacturing Technologies* (Vol. 17). Springer, Cham, Switzerland.
5. Vayre, B., Vignat, F., & Villeneuve, F. (2012). Designing for additive manufacturing. *Procedia CIrP*, 3, 632–637.
6. Lin, Z., Song, K., & Yu, X. (2021). A review on wire and arc additive manufacturing of titanium alloy. *Journal of Manufacturing Processes*, 70, 24–45.
7. Tripathi, U., Saini, N., Mulik, R. S., & Mahapatra, M. M. (2022). Effect of build direction on the microstructure evolution and their mechanical properties using GTAW based wire arc additive manufacturing. *CIRP Journal of Manufacturing Science and Technology*, 37, 103–109.
8. Bourell, D., Kruth, J. P., Leu, M., Levy, G., Rosen, D., Beese, A. M., & Clare, A. (2017). Materials for additive manufacturing. *CIRP Annals*, 66(2), 659–681.
9. Guo, N., & Leu, M. C. (2013). Additive manufacturing: Technology, applications and research needs. *Frontiers of Mechanical Engineering*, 8(3), 215–243.
10. Gao, W., Zhang, Y., Ramanujan, D., et al. (2015). The status, challenges, and future of additive manufacturing in engineering. *Computer-Aided Design*, 69, 65–89.
11. Mok, S. H., Bi, G., Folkes, J. I., & Pashby, I. (2008). Deposition of Ti-6Al-4V using a high power diode laser and wire, part I: Investigation on the process characteristics. *Surface and Coatings Technology*, 202, 3933–3939.
12. Brandl, E., Michailov, V., Viehweger, B., & Leyens, C. (2011). Deposition of Ti-6Al-4V using laser and wire, part I: Microstructural properties of single beads. *Surface and Coatings Technology*, 206, 1120–1129.

144 Additive Manufacturing

13. Baufeld, B., Brandl, E., & van der Biest, O. (2011). Wire based additive layer manufacturing: Comparison of microstructure and mechanical properties of Ti–6Al–4Vcomponentsfabricated by laser-beam deposition and shaped metal deposition. *Journal of Materials Processing Technology*, 211, 1146–1158.

14. Horgar, A., Fostervoll, H., Nyhus, B., Ren, X., Eriksson, M., & Akselsen, O. M. (2018). Additive manufacturing using WAAM with AA5183 wire. *Journal of Materials Processing Technology*, 259, 68–74.

15. Rodrigues, T. A., Duarte, V., Miranda, R. M., Santos, T. G., & Oliveira, J. P. (2019). Current status and perspectives on wire and arc additive manufacturing (WAAM). *Materials*, 12(7), 1121.

16. Vimal, K. E. K., Srinivas, M. N., & Rajak, S. (2021). Wire arc additive manufacturing of aluminium alloys: A review. *Materials Today: Proceedings*, 41, 1139–1145.

17. Ding, D., Pan, Z., Cuiuri, D., & Li, H. (2015). A multi-bead overlapping model for robotic wire and arc additive manufacturing (WAAM). *Robotics and Computer-Integrated Manufacturing*, 31, 101–110.

18. Gierth, M., Henckell, P., Ali, Y., Scholl, J., & Bergmann, J. P. (2020). Wire arc additive manufacturing (WAAM) of aluminum alloy AlMg5Mn with energy-reduced gas metal arc welding (GMAW). *Materials*, 13(12), 2671.

19. Veiga, F., Gil Del Val, A., Suárez, A., & Alonso, U. (2020). Analysis of the machining process of titanium Ti6Al-4V parts manufactured by wire arc additive manufacturing (WAAM). *Materials*, 13(3), 766.

20. Lu, X., Zhou, Y. F., Xing, X. L., Shao, L. Y., Yang, Q. X., & Gao, S. Y. (2017). Open-source wire and arc additive manufacturing system: Formability, microstructures, and mechanical properties. *The International Journal of Advanced Manufacturing Technology*, 93(5), 2145–2154.

21. Rosli, N. A., Alkahari, M. R., bin Abdollah, M. F., Maidin, S., Ramli, F. R., & Herawan, S. G. (2021). Review on effect of heat input for wire arc additive manufacturing process. *Journal of Materials Research and Technology*, 11, 2127–2145.

22. Thapliyal, S. (2019). Challenges associated with the wire arc additive manufacturing (WAAM) of aluminum alloys. *Materials Research Express*, 6(11), 112006.

23. Paskual, A., Álvarez, P., & Suárez, A. (2018). Study on arc welding processes for high deposition rate additive manufacturing. *Procedia Cirp*, 68, 358–362.

24. Shi, J., Li, F., Chen, S., Zhao, Y., & Tian, H. (2019). Effect of in-process active cooling on forming quality and efficiency of tandem GMAW–based additive manufacturing. *The International Journal of Advanced Manufacturing Technology*, 101(5), 1349–1356.

25. Wu, B., Pan, Z., Ding, D., Cuiuri, D., Li, H., Xu, J., & Norrish, J. (2018). A review of the wire arc additive manufacturing of metals: Properties, defects and quality improvement. *Journal of Manufacturing Processes*, 35, 127–139.

26. Wu, B., Pan, Z., Ding, D., Cuiuri, D., Li, H., & Fei, Z. (2018). The effects of forced interpass cooling on the material properties of wire arc additively manufactured Ti6Al4V alloy. *Journal of Materials Processing Technology*, 258, 97–105.

27. Pan, Z., Ding, D., Wu, B., Cuiuri, D., Li, H., & Norrish, J. (2018). Arc welding processes for additive manufacturing: A review. *Transactions on Intelligent Welding Manufacturing*, 3–24.

28. Sequeira Almeida, P. M., & Williams, S. (2010). Innovative process model of Ti-6Al-4V additive layer manufacturing using cold metal transfer (CMT). In *Proceedings of the 21st Annual International Solid Freeform Fabrication Symposium*, Austin, TX, 9–11 August 2010, 25–36.

29. Boyer, B., Welsch, G., & Collings, E. (2007). *Materials Properties Handbook: Titanium Alloys*. ASM International, Materials Park, OH, 790–810.

30. Ferraris, S., & Volpone, L. (2005). Aluminium alloys in third millennium shipbuilding: materials, technologies, perspectives. In *The Fifth International Forum on Aluminium Ships*, Tokyo, Japan, Citeseer.

31. Inagaki, I., Takechi, T., Shirai, Y., & Ariyasu N. (2014). Application and features of titanium for the aerospace industry. *Nippon Steel Sumitomo Metal Technical Report*, 106(106), 22–27.

32. Montgomery, J. S., & Wells, M. G. (2001). Titanium armor applications in combat vehicles. *JOM*, 53(4), 29–32.

33. Li, Z., Liu, C., Xu, T., et al. (2019). Reducing arc heat input and obtaining equiaxed grains by hot-wire method during arc additive manufacturing titanium alloy. *Materials Science and Engineering A*, 742, 287–294.

34. Martina, F., Williams, S. W., & Colegrove, P. (2013). Improved microstructure and increased mechanical properties of additive manufacture produced Ti-6Al-4V by interpass cold rolling. In *2013 International Solid Freeform Fabrication Symposium*. University of Texas, Austin.

35. Zhou, Y., Qin, G., Li, L., Lu, X., Jing, R., Xing, X., & Yang, Q. (2020). Formability, microstructure and mechanical properties of Ti-6Al-4V deposited by wire and arc additive manufacturing with different deposition paths. *Materials Science and Engineering: A*, 772, 138654.

36. Kumar, S., Karpagaraj, A., & Kumar, R. (2020). A review on arc welding of super duplex stainless steel (SDSS) 2507. *Manufacturing Technology Today*, 19(1–2), 17–24.

37. Shen, C., Pan, Z., Cuiuri, D., Roberts, J., & Li, H. (2016). Fabrication of Fe-FeAl functionally graded material using the wire-arc additive manufacturing process. *Metallurgical and Materials Transactions B*, 47(1), 763–772.

38. Zhuo, Y., Yang, C., Fan, C., Lin, S., Chen, C., & Cai, X. (2020). Microstructure and mechanical properties of wire arc additive repairing Ti–5Al–2Sn–2Zr–4Mo–4Cr titanium alloy. *Materials Science and Technology*, 36(15), 1712–1719.

39. Wu, B., Ding, D., Pan, Z., et al. (2017). Effects of heat accumulation on the arc characteristics and metal transfer behavior in wire arc additive manufacturing of Ti6Al4V. *Journal of Materials Processing Technology*, 250, 304–312.

40. Candel-Ruiz, A., Kaufmann, S., & Müllerschön, O. (2015). Strategies for high deposition rate additive manufacturing by laser metal deposition. In *Proceedings of Lasers in Manufacturing (LiM)*. German Scientific Laser Society (WLT e.V.), Erlangen, Germany, 584–610.

41. Wu, B., Pan, Z., Ding, D., Cuiuri, D., & Li, H. (2018). Effects of heat accumulation on microstructure and mechanical properties of Ti6Al4V alloy deposited by wire arc additive manufacturing. *Additive Manufacturing*, 23, 151–160.

42. Wang, D., Li, H., Wang, X., Zheng, W., Lin, Z., & Liu, G. (2019). The microstructure evolution and mechanical properties of TiBw/TA15 composite with network structure prepared by rapid current assisted sintering. *Metals*, 9(5), 540.

43. Lin, J. J., Lv, Y. H., Liu, Y. X., et al. (2016). Microstructural evolution and mechanical properties of Ti-6Al-4V wall deposited by pulsed plasma arc additive manufacturing. *Materials Design*, 102, 30–40.

44. Xu, F., Madhaven, N., Dhokia, V., et al. (2016). Multi-sensor system for wire-fed additive manufacture of titanium alloys. In *26th International Conference on Flexible Automation and Intelligent Manufacturing*.

45. Panchenko, O., Kurushkin, D., Mushnikov, I., Khismatullin, A., & Popovich, A. (2020). A high-performance WAAM process for Al–Mg–Mn using controlled short-circuiting metal transfer at increased wire feed rate and increased travel speed. *Materials & Design*, 195, 109040.

46. Gou, J., Shen, J., Hu, S., Tian, Y., & Liang, Y. (2019). Microstructure and mechanical properties of as-built and heat-treated Ti-6Al-4V alloy prepared by cold metal transfer additive manufacturing. *Journal of Manufacturing Processes*, 42, 41–50.

47. Shunmugavel, M., Polishetty, A., & Littlefair, G. (2015). Microstructure and mechanical properties of wrought and additive manufactured Ti-6Al-4V cylindrical bars. *Procedia Technology*, 20, 231–236.

48. Al-Bermani, S. S., Blackmore, M. L., Zhang, W., & Todd, I. (2010). The origin of microstructural diversity, texture, and mechanical properties in electron beam melted Ti-6Al-4V. *Metallurgical and Materials Transactions A*, 41(13), 3422–3434.

49. Vazquez, L., Rodriguez, M. N., Rodriguez, I., & Alvarez, P. (2021). Influence of post-deposition heat treatments on the microstructure and tensile properties of Ti-6Al-4V parts manufactured by CMT-WAAM. *Metals*, 11(8), 1161.

50. Antonysamy, A. A. (2012). *Microstructure, Texture and Mechanical Property Evolution During Additive Manufacturing of Ti6Al4V Alloy for Aerospace Applications*. The University of Manchester, United Kingdom.

51. Zhang, P. L., Jia, Z. Y., Yan, H., et al. (2021). Effect of deposition rate on microstructure and mechanical properties of wire arc additive manufacturing of Ti-6Al-4V components. *Journal of Central South University*, 28(4), 1100–1110.

52. Donoghue, J., Antonysamy, A. A., Martina, F., Colegrove, P. A., Williams, S. W., & Prangnell, P. B. (2016). The effectiveness of combining rolling deformation with wire–arc additive manufacture on β-grain refinement and texture modification in Ti–6Al–4V. *Materials Characterization*, 114, 103–114.

53. Chi, J., Cai, Z., Wan, Z., et al. (2020). Effects of heat treatment combined with laser shock peening on wire and arc additive manufactured Ti17 titanium alloy: Microstructures, residual stress and mechanical properties. *Surface and Coatings Technology*, 396, 125908.

54. Li, K., Klecka, M. A., Chen, S., & Xiong, W. (2021). Wire-arc additive manufacturing and post-heat treatment optimization on microstructure and mechanical properties of Grade 91 steel. *Additive Manufacturing*, 37, 101734.

55. Wilson-Heid, A. E., Wang, Z., McCornac, B., & Beese, A. M. (2017). Quantitative relationship between anisotropic strain to failure and grain morphology in additively manufactured Ti-6Al-4V. *Materials Science and Engineering A*, 706, 287–294.

56. Gou, J., Wang, Z., Hu, S., Shen, J., Tian, Y., Zhao, G., & Chen, Y. (2020). Effects of trace Nb addition on microstructure and properties of Ti–6Al–4V thin-wall structure prepared via cold metal transfer additive manufacturing. *Journal of Alloys and Compounds*, 829, 154481.

57. Liberini, M., Astarita, A., Campatelli, G., Scippa, A., Montevecchi, F., Venturini, G., Durante, M., Boccarusso, L., Minutolo, F. M. C., & Squillace, A. (2017). Selection of optimal process parameters for wire arc additive manufacturing. *Procedia CIRP*, 62, 470–474. https://doi.org/10.1016/j.procir.2016.06.124.

58. Yang, D., He, C., & Zhang, G. (2016). Forming characteristics of thin-wall steel parts by double electrode GMAW based additive manufacturing. *Journal of Materials Processing Technology*, 227, 153–160. https://doi.org/10.1016/j.jmatprotec.2015.08.021.

59. Masubuchi, K. (2013). *Analysis of Welded Structures: Residual Stresses, Distortion, and Their Consequences*. Elsevier.

60. Colegrove, P. A., Coules, H. E., Fairman, J., et al. (2013). Microstructure and residual stress improvement in wire and arc additively manufactured parts through high-pressure rolling. *Journal of Materials Processing Technology*, 213, 1782–1791.

61. Hoye, N. (2015). Characterisation of Ti-6Al-4V deposits produced by arc-wire based additive manufacture. Dissertation, University of Wollongong.

62. Peyre, P., Fabbro, R., Merrien, P., & Lieurade, H. P. (1996). Laser shock processing of aluminium alloys. Application to high cycle fatigue behaviour. *Materials Science and Engineering: A*, 210(1–2), 102–113.

63. Antunes, R. A., & de Oliveira, M. C. (2020). Stress corrosion cracking of structural nuclear materials: Influencing factors and materials selection. *Innovations in Corrosion and Materials Science (Formerly Recent Patents on Corrosion Science)*, 10(1), 5–24.

64. Cheng, X., Fisher, J. W., Prask, H. J., et al. (2003). Residual stress modification by post-weld treatment and its beneficial effect on fatigue strength of welded structures. *International Journal of Fatigue*, 25, 1259–1269.

65. Wu, B., Pan, Z., Ding, D., Cuiuri, D., Li, H., & Fei, Z. (2018). The effects of forced interpass cooling on the material properties of wire arc additively manufactured Ti6Al4V alloy. *Journal of Materials Processing Technology*, 258, 97–105.

66. Zhuo, Y., Yang, C., Fan, C., & Lin, S. (2021). Grain morphology evolution mechanism of titanium alloy by the combination of pulsed arc and solution element during wire arc additive manufacturing. *Journal of Alloys and Compounds*, 888, 161641.

67. Wu, B., Pan, Z., Ding, D., Cuiuri, D., Li, H., & Fei, Z. (2018). The effects of forced interpass cooling on the material properties of wire arc additively manufactured Ti6Al4V alloy. *Journal of Materials Processing Technology*, 258, 97–105.

68. Pan, Z, Cuiuri, D, Roberts, J., et al. (2016). Fabrication of Fe-FeAl functionally graded material using the wire-arc additive manufacturing process. *Metallurgical and Materials Transactions B*, 47, 763–772.

69. Martina, F., Colegrove, P. A., Williams, S. W., & Meyer, J. (2015). Microstructure of interpass rolled wire+ arc additive manufacturing Ti-6Al-4V components. *Metallurgical and Materials Transactions A*, 46, 6103–6118.

70. Wang, F., Williams, S., Colegrove, P., & Antonysamy, A. A. (2013). Microstructure and mechanical properties of wire and arc additive manufactured Ti-6Al-4V. *Metallurgical and Materials Transactions A*, 44(2), 968–977.

Chapter 9

Advances in additive manufacturing

M.B. Kiran and V.J. Badheka
Pandit Deendayal Energy University

CONTENTS

9.1	Introduction	150
9.2	Recent developments in additive manufacturing	150
	9.2.1 AM processes	150
	9.2.1.1 Laser beam melting (LBM)	150
	9.2.1.2 Electron beam melting (EBM)	152
	9.2.1.3 Laser metal deposition (LMD)	152
	9.2.2 Microstructure and properties	153
	9.2.2.1 Microstructure of laser beam melting	153
	9.2.2.2 Microstructure of electron beam melting	154
	9.2.2.3 Microstructure laser metal deposition	156
9.3	Application of artificial intelligence in additive manufacturing	157
	9.3.1 Printability	157
	9.3.2 Efficiency in pre-fabrication	157
	9.3.3 Service-oriented architecture (SOA)	158
	9.3.4 Defect classification	158
	9.3.4.1 Types of defects	158
	9.3.4.2 Product inspection by using artificial neural networks	159
	9.3.4.3 Real-time build control	159
	9.3.4.4 Predictive maintenance	160
	9.3.5 Waste reduction	160
	9.3.6 Reducing energy consumption	161
9.4	Applications of additive manufacturing	161
	9.4.1 Surgical implants	161
	9.4.2 Sensors	161
	9.4.3 Topology optimization	162
9.5	Future directions	162
References		163

DOI: 10.1201/9781003258391-9

9.1 INTRODUCTION

In conventional manufacturing processes, such as turning, milling, shaping, etc., a component is made by removing material, from a workpiece, in the form of chips. The chips will be generated when a cutting tool is forced into the workpiece. The selection of cutting tools used depends upon the type of manufacturing process. For example, in straight turning, a single-point tool is used. Thus, every traditional manufacturing process uses a device while shaping the component. During the manufacturing of a component, a lot of material will be wasted in the form of chips. The traditional techniques are capable of producing products with high quality and finish. However, these processes cannot produce components with complicated geometries. Researchers have proposed a new additive manufacturing (AM) technique to overcome the drawbacks of traditional manufacturing processes. Using computer aided design (CAD) data, a product is constructed by depositing material one layer after another. It is called AM/or 3D printing. Unlike traditional machining, AM does not use any tool while shaping a product. Also, there will be minimum wastage of material. AM is attracting a large number of researchers because of its scope. The authors identified practitioners, researchers, and academicians who require comprehensive review material on the evolution of AM. This has motivated the authors to work on this review project (Figure 9.1).

9.2 RECENT DEVELOPMENTS IN ADDITIVE MANUFACTURING

An attempt has been made, in this section, to discuss (i) AM processes and (ii) using artificial intelligence (AI) in AM.

9.2.1 AM processes

9.2.1.1 Laser beam melting (LBM)

During this process, metal powder is spread evenly over a metallic platform. A leveling system is used for evenly distributing the metal powder to a lamellar thickness of about 50–100 μm. The platform has a size of 50×50 mm (Tan et al. 2016). The energy used for melting the metal powder is supplied by a laser beam, with a λ of 1060 nm (Dai et al. 2016). 20 Watts to 1 KW power is consumed by the laser beam (Zhao et al. 2016). The laser beam will have a size of around 50 μm and will be scanning the metal powder at 16 m/s. The metal powder will be melted because of the intense heat. The metal in a liquid state will fuse with the already solidified metal. After the liquid metal solidifies, a layer is completed, and the build plate is made to go down. The above steps are repeated until the

Advances in additive manufacturing 151

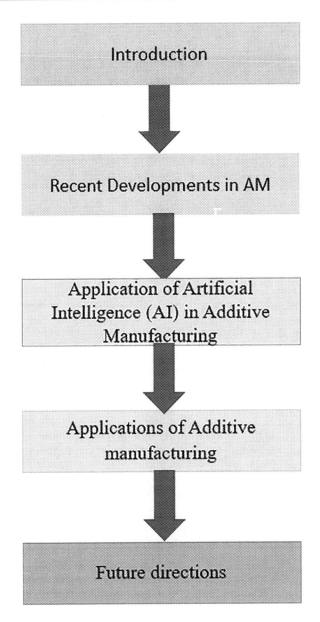

Figure 9.1 Flowchart of arrangement of topics.

entire object becomes ready. The LBM process is executed in an enclosure. The enclosure contains Argon/or Nitrogen. Figure 9.2 shows the working principle of LBM.

LBM has no moving components, and hence the process is capable of producing 3D objects with detailed profiles.

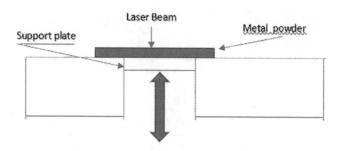

Figure 9.2 Working principle of laser beam melting.

9.2.1.2 Electron beam melting (EBM)

In this method, a beam of electrons acts as the primary energy source. Sixty kilovolts of power is used for accelerating the electrons. The electron beam produces 700°C for melting powder (Chen et al. 2017), which is spread evenly on the plate by using a leveling system. Thirty milliamperes are used for producing an electron beam with a size of 50 μm and a wavelength of 1060 nm (Dai et al. 2017). The speed of the electron beam, while scanning, is 104 mm/s (Dai et al. 2016). The heat from the electron beam during scanning is used for melting the metal powder. The liquid metal will mix with the solidified metal. The build plate is moved down after solidification. The above steps will be repeated till the final product becomes ready. The chamber of the EBM machine has a vacuum lower than 10^{-2} Pa. This is done to give protection to the molten metal. Figure 9.3 shows the working principle.

Specific challenges need to be met to get quality products from the EBM process. Designers should know the geometry of the part. In addition, an understanding of overhang requirements, part inclination, the finish requirements expected on the surface, information on tolerance, post-processing requirements is required for producing quality products.

Any product made by EBM requires post-processing. The post-processing operation will help in removing the loose powder. Post-processing of EBM-made products is usually done by sandblasting. Also, the products made by EBM will have residual stresses. The presence of these stresses is harmful to the effective functioning of the product. These residual stresses are removed by heat treatment operation—the everyday heat treatment operations are hot isostatic pressing, vacuum heat treatment process, etc.

9.2.1.3 Laser metal deposition (LMD)

A nozzle is used for feeding metal powder (Zhang et al. 2011). The metal powder is melted using the heat obtained from the CO_2 laser. The nozzle is specially designed so that CO_2 enters from the center of the nozzle. Figure 9.3 shows the experimental setup. Argon/or Helium prevents the molten

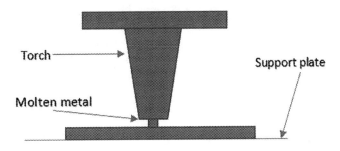

Figure 9.3 Working principle of electron beam melting.

metal's oxidation (Grotowski et al. 2010). LMD can produce higher build volumes (Bai et al. 2017). Damaged components is repaired by using the LMD process. Liquid metal particles are deposited directly using a laser beam while producing a 3D object. Material wastage is less in the case of the LMD process.

Damaged products are repaired extensively using the LMD process. In the LMD process, the metal products are liquefied and are deposited while producing a 3D product. The LMD process is very efficient as there is minimum wastage. Both LBM and LMD make use of an energy source. The LMD process cannot be used while manufacturing products having complicated shapes. The LBM process is preferred when a product with a complicated profile is required. The main drawback of AM is that it consumes very high power. Another challenge is to reduce the wastage regarding supporting plates. The other challenges include surface roughness enhancement of manufactured products, hazards related to health, selection of optimum process parameters, and enhancing the mechanical properties of manufactured products. This is because products made by additive manufactured products do not have the required density, resulting in reduced mechanical properties. On the other hand, products made by the traditional manufacturing process will have increased density and hence good mechanical properties. It was also observed that AM releases harmful small particles every minute. These will have a detrimental effect on the health of human beings. The application of AI in AM makes it possible to improve the productivity of AM process.

9.2.2 Microstructure and properties

9.2.2.1 Microstructure of laser beam melting

The microstructure obtained in AM will depend on the scan speed (Attar et al. 2014). For example, coarse grains will form at 100 mm/s and use 90 W supply power. The formation of coarse grain during cooling is due to the transformation of the β particle into the α particle. The literature also

154 Additive Manufacturing

reported that the martensitic α phase would be created if a scanning speed >200 mm/s is used. During AM, quick cooling causes huge thermal gradients. This would result in the formation of significant residual stresses. This is detrimental to the properties of the manufactured product. For example, the inclusion of residual stresses would reduce the manufactured product's strength.

9.2.2.2 Microstructure of electron beam melting

It was observed that a product made by using EBM is found to consist of β particles with clearly defined grain borders and α–β microstructural arrangement. It was also found that the products made by EBM and Selective laser melting (SLM) processes will have entirely different microstructures (Figure 9.4). High cooling rates during the process convert phases of β into α-martensite.

To increase the reliability of EBM-made products, one has to exercise control during manufacturing. By exercising proper power, mechanical properties can be controlled. The EBM process has a unique property in that it has many micro-sized melting zones. The arrangement or structure of these small melting regions will dictate the properties of the EBM-made product. Thus, to enhance the dependability of the EBM-made product, one has to exercise control over these small melting regions. Researchers observed that Ultimate tensile strength (UTS) and elongation of EBM-made products are not only stable but are also more reliable. Researchers observed that the thickness of β grains largely depends upon the height of the manufactured product. This is due to the prolonged cooling effect with increased product height. Tan et al. (2016) have found that the microstructures of the first layer were different from that of the last layer. They have also observed that the product's strength increased by reducing lamellae thickness. They have also observed that increasing the grain size would reduce lamellae thickness. They have concluded that lowering the lamellae thickness resulted in the increased tensile strength of the manufactured product. When an EBM-made product is heated above β, transit temperature produces a microstructure similar to that obtained from annealing treatment. It was also observed that EBM-made products are more potent than that produced from forging.

The tensile strength of EBM-made products is primarily governed by the diameter of the component (Zhao et al. 2016). Their investigation also revealed that it would enhance tensile strength when the product's diameter increases. The EBM-made part will have a microstructure composed of different phases. It exhibited less corrosion resistance when a microstructure had an α-phase arranged in the Hexagonal close-packed (HCP) structure (Chen et al. 2017). When a microstructure has a -phase arrangement in a Body centered cell (BCC) structure, it has increased corrosion resistance (Todai et al. 2016). During the EBM process, the α phase will start growing over the existing β grains. The size of these β and α particles depends upon the cooling rate.

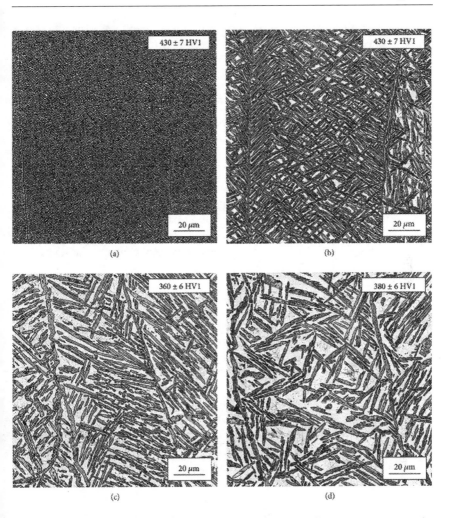

Figure 9.4 (a and b) Microstructure of the specimens at different scanning speeds. (Adapted from Tomasz Kurzynowski (2019) [Open access].) (c and d) Microstructure of the specimens at different scanning speeds. (Adapted from Tomasz Kurzynowski (2019) [Open access].)

It was observed that AM-made titanium alloys exhibit better (Murr et al. 2010) corrosion resistance when compared to wrought products. This is due to the presence (Zhang et al. 2011) of not only the β particles but also the presence of fine α/β particles (Bai et al. 2017). Many of the AM products made of titanium sometimes (Grotowski et al. 2010) are found to consist of pores in their microstructure. These titanium alloys will also have good mechanical properties. For this reason, they are being increasingly used in health industries. It was found that products made by EBM and SLM processes will have different microstructures. The properties of AM products,

156 Additive Manufacturing

like strength, etc., can be improved by heat treatment. AM products are found to have defects, as explained in the following paragraphs.

It was reported in the literature that porous structures in AM products are mainly due to improper selection of process parameters. It was observed that the ductility of AM products depends upon both pore geometry and orientation (Vilaro et al. 2011). Pores become epicenters during crack propagation (Biswas et al. 2012). The diameter of pores present in AM products may be as big as 300 µm. It was found that the lamellae thickness and scanning speed would determine the quality of AM products. Insufficient use of energy during AM process would result in a defect called lack of fusion. This type of defect, when present in AM products, would result in fatigue failure. On the other hand, usage of excess energy during AM process would result in a defect called Keyhole collapse. The lack of dampening of solid particles would result in a Balling defect. The AM process would result in a rough surface due to the layer-by-layer formation of a 3D object.

9.2.2.3 Microstructure laser metal deposition

LMD-processed alpha+beta titanium alloy products will have primary grains made of beta grains. The boundaries of these alloys will have alpha grains at the grain boundaries (Van Arkel 2015). They have excellent fatigue strength. These alloys will have very poor flexibility.

In the case of duplex microstructure (Morita et al. 2005), particles of alpha are dispersed in the matrix of beta particles. They have excellent properties such as tensile strength (Mardaras et al. 2017). In the case of the Basket-weave microstructure, the microstructure will have alpha particles. These alloys will have excellent elasticity. In the case of the equiaxed structure, the microstructure was found to consist of alpha and beta grains. These grains will have a polygonal shape. By heat treating, AM products' desired mechanical properties are obtained.

Solution heat treatment and aging improved mechanical properties (Leyens and Peters 2003). They have achieved increased ductility by using this method. Researchers have demonstrated that the microstructure largely determines the mechanical properties. They also showed that ductility of these alloys would be improved by stress relief annealing. This heat treatment would produce alpha particles with a coarser size.

Scientists have also demonstrated that tensile properties depend on the cooling rates used during the processes. After proper heat treatment, tensile strength and the balance of flexibility would improve significantly. The deposition rates obtained in LMD may be as high as 0.5 kg/h. Thus, for higher productivity, higher deposition rates are required.

Very little research is reported on shielding done during LMD. This would significantly reduce manufacturing costs. Using proper process parameters, the strength obtained by AM is comparable to that obtained by traditional manufacturing processes. Researchers have also demonstrated that the static and fatigue strength of additively manufactured products depend on the microstructure. The following paragraphs explain how the productivity of AM process can be improved by employing AI.

9.3 APPLICATION OF ARTIFICIAL INTELLIGENCE IN ADDITIVE MANUFACTURING

9.3.1 Printability

Certain products can be easily made by AM. These products have good printability (Telea and Jalba 2011). Any product can be made by AM, in principle. Researchers have shown that the applicability of AM is limited to the component geometry and the type of material. The AM process will be considered only when customer requirements can be met. Using this algorithm, one can determine whether a product can be made by AM or not.

The algorithm has different modules. Module 1 is used for extracting the features of the product. The module uses machine learning for removing the elements of an object. Module 2 is used for assessing printability by looking into cost considerations.

9.3.2 Efficiency in pre-fabrication

Today's customer satisfaction requires making products having complicated shapes and sizes. They also need products having complex geometrical features. Manufacturing a product by 3D printing, having complex geometries requires more time. Much of this time is consumed in slicing the 3D data. Slicing is generating a tool path for the printer head. This is analogous to tool position data in the machining of the component using a Computer numerical control (CNC) machine. As in the case of the CNC machine tool, the cutter data are used by the cutting tool for proper movement during machining. Similarly, the slicing data will be used by the printer head in manufacturing the required 3D object. The slicing process is improved by employing AI (Kulkarni et al. 2000).

It was reported in the literature that the adaptive slicing algorithm designed and developed by Gregori et al. (2014) was efficient in terms of using computational resources. Much more work is needed by researchers to make these slicing algorithms computationally efficient. Much more work is required on using parallel computing for improving the efficiency of slicing algorithms. The advent of Big data and Industry 4.0 necessitates parallel computing as these technologies require more computational resources. Literature has reported using dual processors for improving the pre-fabrication efficiency in AM.

9.3.3 Service-oriented architecture (SOA)

Researchers have started designing algorithms based on SOA while performing AM processes (Hassan 2011, Mell and Grance 2009, Hassan 2009, Tao et al. 2011, Da Silveira et al. 2001). Thus, companies have started implementing AM as a service. For manufacturing, the necessary instructions for 3D printing were given through cloud infrastructure from one end, and these instructions will be executed and the 3D object made at the other end of the cloud. This would be useful in providing desired flexibility in terms of volume and variety. Thus, 3D printing can be started and stopped on a need basis. The main advantage of using cloud-based infrastructure is that different companies can work collaboratively. The cloud-based infrastructure is made secure by using cyber security algorithms. Thus, the cyber-physical system can be protected from intruders and cyberattacks.

9.3.4 Defect classification

Researchers have identified different types of defects in products made using AM.

9.3.4.1 Types of defects

Porosity (Figure 9.5) is a defect found in additive manufactured products. Porosity would be detrimental to the satisfactory functioning of the AM-made products. It was reported in the literature that AM-made products might have a porosity as high as 1%–5%.

The porosity in AM products would lead to product failure. Research has shown that the pores' shape, size, and geometry affect the product's functioning and ultimately lead to product failure. Researchers have found

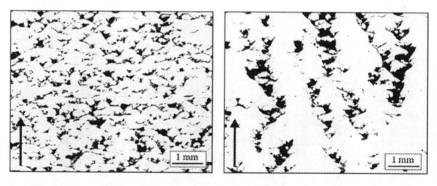

Figure 9.5 Porosity defects. (Adapted from Tomasz Kurzynowski (2019) [Open access].)

that the improper use of process variables would result in porosity-related defects. It was found (Yadroitsev et al. 2007) that hatch distance, when optimized, would minimize the porosity-related defects in AM-made products. In the experiment, they used a 70 μm diameter laser and a 120 μm hatch width for minimum porosity. The relationship between hatch distance and porosity was also studied by other researchers (Mireles et al. 2015). Gas entrapment during the making of products by AM would result in gas pores. Studies have shown that it is impossible to do away with gas pores in AM (Ng et al. 2009). Investigations revealed that AM-made products might have as high as 0.7% gas pores. It was also observed that the gas pores in AM products would act as initiators in crack propagation (Sercombe et al. 2008).

9.3.4.2 Product inspection by using artificial neural networks

Many researchers have tried to mimic the human brain, resulting in the emergence of artificial neural networks (ANN). In ANN, there will be input nodes used to receive data from external devices. The primary function of the input layer is in calculating weights and passing them to the processing (hidden) layer. The number of processing layer nodes depends upon the type of problem. This layer knows how to process weights received from the input layer and then sends the information to the output layer. Depending upon the type/complexity of the problem, nodes will be selected for the output layer. This layer would help in sending the transformed information to the end-user.

During the training of the ANN, the input layer is made to calculate the set of optimum weights. During the validation phase, the output of the ANN is compared with the expected production called supervised learning. During the training of an ANN, for the first iteration, weights are selected randomly. This will ensure a minor deviation between the expected and actual results. This way, many iterations are made.

During training an ANN, both the inputs and outputs are made known to the network. Network training time would depend on the complexity of the problem. When the work from a network reaches the expected performance, the training is said to be complete. At this level, the activity may be stopped. Now the network is ready to inspect the test data set. When the network gives an expected performance at this stage, the ANN has understood the general patterns in the problem domain.

9.3.4.3 Real-time build control

The quality of AM-manufactured product depends on scan speed, layer thickness, and quantity of the molten metal. In this way, the quality of AM products can be controlled in real-time. Machine learning is extensively used in real-time manufacturing control (Edward et al. 2018). The inputs required for performing real-time build control are component geometry, a data set for training, executing free form deposition.

9.3.4.4 Predictive maintenance

Routine maintenance of a machine tool would enhance its service life. A schedule is established, and then the preventive maintenance (PM) is executed as per the predefined schedule. In contrast, repair will be done after machine failure in the case of breakdown performance. With the availability of sensors, researchers have proposed new ways of maintaining machines cost-effectively. This is called condition-based maintenance. In these systems, the sensors capture the online data on process parameters. Knowledge-based models will use these data for decision-making. The models are used for assessing the service life of machines. In this way, AI is being used to prevent machine breakdowns. Also, this will help in increasing the availability as well as the quality of manufactured products. Condition-based maintenance of machines attracts many researchers (Yam et al. 2001). They were also successful in fault diagnosis and the assessment of machine deterioration. This helps in choosing the right maintenance policy for a given machine tool.

For the PM of machine tools, cluster identification is a practical step after data collection. Researchers have identified critical steps in identifying data clusters (Jardine et al. 2006). They have used this technique for the PM of the SLM machine.

9.3.5 Waste reduction

A component is made in AM layer by layer. Hence, 3D printing cannot be used to manufacture parts with overhangs (Leary et al. 2014). To overcome this drawback, proper support is provided for components having an overhang. The main problem in using the supports is that these supports are no longer required after manufacturing. This is a non-value-adding activity. This would increase the manufacturing cost. Solving part orientation problems is attracting many researchers. This would reduce the support requirements and hence the cost of production. This would be a significant saving (Pham et al. 1999, Frank and Fadel 1995, Strano et al. 2013, Das et al. 2017, Morgan et al. 2016). Researchers explored using relatively cheap materials for making support. After the manufacturing gets over, the support material would be dissolved in one scheme. Acrylic copolymers are used as support material (Hopkins et al. 2009). In this way, support materials were removed quickly. Acrylonitrile butadiene styrene material is also being used as a supporting material with a painting by polylactic acid (Domonoky et al. 2016). After manufacturing a 3D object, the support material is removed using isopropyl alcohol and potassium hydroxide. Ni et al. (2017) investigated using polyvinyl alcohol as a support material in the AM process.

9.3.6 Reducing energy consumption

AM consumes a high amount of energy when compared to traditional manufacturing. There is a huge need for reducing the energy consumption by the AM process. Selective laser melting sintering (SLS) process makes use of laser for getting power. The powder is spread evenly on a metal plate. During 3D printing operations, laser light is used for melting the metal powder. In this way, the first layer is formed. Again, a new lamella of metal particles is spread evenly using a roller. The above steps are repeated till the finished product is ready. Thus, in 3D printing, the product is made layer by layer. The SLS process was investigated by many researchers (Mansour and Hague 2003) by using various types of polymers. The energy supplied will be used for processing and other supporting functions in the SLS process. Investing activities are moving pistons, re-coater arm, and heating. It was reported in the literature that 56% of the energy is consumed for processing. Thus, a large proportion of energy is consumed for performing supporting activities. Therefore, there is a vast potential for improving energy efficiency while performing supporting activities. This will also make the AM process more economical. The absorptivity of material, laser intensity, laser speed, and the size of the laserspot will decide the amount of energy consumed in the AM process (Lu 2016).

Thus, there is a huge potential in transforming AM processes environment-friendly.

9.4 APPLICATIONS OF ADDITIVE MANUFACTURING

The following paragraphs highlight the applications of AM.

9.4.1 Surgical implants

Metallic elements were earlier used to make surgical implants. This is because the metallic elements have good bearing properties. Many surgical implants are made using metal and polymer combinations with AM. The traditional manufacturing processes could not handle the combination of metal and polymer.

9.4.2 Sensors

Strain sensors are used for the measurement of tensile strain and compressive strain. During the height, tensile stress or compressive strains will be converted into electrical signals. The Fuse diffusion modeling (FDM)

162 Additive Manufacturing

Figure 9.6 Topology optimization.

technique is commonly used for manufacturing these sensors. It was claimed by Mohammad Reza Khosravani et al. that these sensors could withstand strains up to 800%.

9.4.3 Topology optimization

Gyroids are used in solar cells (Figure 9.6). Topology optimization is done by keeping various unit cell types in the gyroid structure to increase the system's surface area for better heat transfer. The component was made by Pandit Deendayal Energy University facility, using a laser metal printer and SS316L material.

9.5 FUTURE DIRECTIONS

Though many researchers have been contributing to the field of AM, many challenges still need to be addressed. The following are the items that require immediate attention by researchers:

    a. To reduce energy consumption by AM processes
    b. To enhance the safety and well-being of people working with an AM process

c. To enhance the security of AM infrastructure from cyberattacks
d. To enhance the mechanical properties of AM-made products
e. To enhance the capability of AM processes, such as to improve surface finish
f. To reduce material waste
g. Additive friction stir deposition is attracting many researchers who are based on solid-state friction stir welding (FSW)

REFERENCES

Attar, H. et al. Comparative study of microstructures and mechanical properties of in situ Ti–TiB composites produced by selective laser melting, powder metallurgy, and casting technologies. *Journal of Materials Research*, 29, 1941–1950, 2014. https://doi.org/10.1557/jmr.2014.122.

Bai, Y., Gai, X., Li, S., Zhang, L-C., Liu, Y., Hao, Y., Zhang, X., Yang, R., and Gao, Y. Improved corrosion behavior of electron beam melted Ti-6Al–4V alloy in phosphate buffered saline. *Corrosion Science*, 123, 289–296, 2017. ISSN 0010–938X, https://doi.org/10.1016/j.corsci.2017.05.003.

Biswas, N., Ding, J.L., Balla, V.K., Field, D.P., and Bandyopadhyay, A. Deformation, and fracture behavior of laser processed dense and porous Ti6Al4V alloy under static and dynamic loading. *Materials Science and Engineering: A*, 549, 13–221, 2012.

Chen, Y., Zhang, J., Dai, N., Qin, P., Attar, H., and Zhang, L.C. The corrosion behavior of selective laser melted Ti-TiB bio-composite in simulated body fluid. *Electrochimica Acta*, 232, 89–97, 2017.

Da Silveira, G., Borenstein, D., and Fogliatto, H.S. Mass customization: Literature review and research directions. *International Journal of Production Economics*, 72, no. 49, 1–13, 2001.

Dai, N., Zhang, L.C., Zhang, J., Chen, Q., and Wu, M. Corrosion behavior of selective laser melted Ti-6Al-4V alloy in NaCl solution. *Corrosion Science*, 102, 484–489, 2016.

Das, P., Mhapsekar, K., Chowdhury, S., Samant, R., and Anand, S., Selection of build orientation for optimal support structures and minimum part errors in additive manufacturing. *Computer-Aided Design, and Applications*, 1–13, 2017.

Domonoky, and Bonsai Brain, Support – Full Disclosure [WWW Document], 2016.

Edward, M.E.H.R., Ellis, T., and Noone, J. Real-time adaptive control of additive manufacturing processes using machine learning, US20180341248A1, 2018.

Frank, D., and Fadel, G. Expert system-based selection of the preferred direction of build for rapid prototyping processes. *Journal of Intelligent Manufacturing*, 6, 339–345, 1995.

Grotowski, V.J.F., Zhang, L.C., and Sercombe, T.B. Prototypes for bone implant scaffolds designed via topology optimization and manufactured by solid freeform fabrication. *Advanced Engineering Materials*, 12, 1106–1110, 2010.

Gregori, R. M. M. H., Volpato, N., Minetto R., and Silva M. V. G. D. Slicing Triangle Meshes: An Asymptotically Optimal Algorithm. *International Conference on Computational Science and ITS Applications*, pp. 252–255, 2014.

Hassan, Q.F. Aspects of SOA: An entry point for starters. *Annals of Computer Science Series*, 7, no. 2, 125, 2009.

Hassan, Q.F. Demystifying cloud security. *CrossTalk*, 16–21, 2011.

Hopkins, P.E., Priedeman Jr. W.R., and Bye, J.F. Support material for digital manufacturing systems, US8246888 B2, 2009.

Jardine, A.K.S., Lin, D., and Banjevic, D. A review on machinery diagnostics and prognostics implementing condition-based maintenance. *Mechanical Systems and Signal Processing*, 20, 1483–1510, 2006.

Kulkarni, P., Marsan, A., and Dutta, D. Review of process planning techniques in layered manufacturing. *Rapid Prototyping Journal*, 6, 18–35, 2000. doi: 10.1108/13552540010309859.

Kurzynowski, T., Madeja, M., Dziedzic, R., & Kobiela, K. The effect of EBM process parameters on porosity and microstructure of Ti-5Al-5Mo-5V-1Cr-1Fe alloy. *Scanning*, 2019, Article ID 2903920, 12 pages. https://doi.org/10.1155/2019/2903920.

Leary, M., Merli, L., Torti, F., Mazur, M., and Brandt, M. Optimal topology for additive manufacture: A method for enabling additive manufacture of support-free optimal structures. *Materials and Design*, 63, 678–690, 2014.

Leyens, C., and Peters, M. *Titanium and Titanium Alloys: Fundamentals and Applications.* Wiley-VCH: Weinheim, Germany; John Wiley: Chichester, UK, 2003.

Mansour, S., and Hague, R. Impact of rapid manufacturing on design for manufacture for injection molding. *Proceedings of the Institution of Mechanical Engineers, Part B: Journal of Engineering Manufacture*, 217, no. 4, 453–461, 2003.

Mardaras, J., Emile, P., and Santgerma, A. Airbus approach for F&DT stress justification of Additive Manufacturing parts. *Procedia Structural Integrity*, 7, 109–115, 2017.

Mell, P., and Grance, T. The NIST definition of cloud computing. *National Institute of Standards Technology*, 53, 50–62, 2009.

Mireles, J., Ridwan, S., Morton, P.A., Hinojos, A., and Wicker, R.B. Analysis, and correction of defects within parts fabricated using powder bed fusion technology. *Surface Topography: Metrology and Properties*, 3, no. 3, 034002, 2015.

Morgan, H.D., Cherry, J.A., Jonnalagadda, S., Ewing, D., and Sienz, J. Part orientation optimisation for the additive layer manufacture of metal components. *The International Journal of Advanced Manufacturing Technology*, 86, 1679–1687, 2016.

Morita, T., Hatsuoka, K., Iizuka, T., and Kawasaki, K. Strengthening of Ti–6Al–4V alloy by short-time duplex heat treatment, *Material Transactions*, 46, 1681–1686, 2005.

Ng, G.K.L., Jarfors, A.E.W., Bi, G., and Zheng, H.Y. Porosity formation and gas bubble retention in laser metal deposition. *Applied Physics A*, 97, no. 3, 641–649, 2009.

Ni, F., Wang, G., and Zhao, H., Fabrication of water-soluble poly (vinyl alcohol)-based composites with improved thermal behavior for potential three-dimensional printing application. *Journal of Applied Polymer Science*, 134, 2017.

Pham, D.T., Dimov, S.S., and Gault, R.S. Part orientation in stereolithography. *International Journal of Advanced Manufacturing Technology*, 15, 674–682, 1999.

Sercombe, T., Jones, N., Day, R., and Kop, A. Heat treatment of Ti-6Al-7Nb components produced by selective laser melting. *Rapid Prototyping Journal*, 14, no. 5, 300–304, 2008.

Strano, G., Hao, L., Everson, R.M., and Evans, K.E. A new approach to the design and optimisation of support structures in additive manufacturing. *International Journal of Advanced Manufacturing Technology*, 66, 1247–1254, 2013.

Tan, X., Kok, Y., Toh, W.Q., Tan, Y.J., Descoins, M., Mangelinck, D., Tor, S.B., Leong, K.F., and Chua, C.K. Revealing martensitic transformation and α/β interface evolution in electron beam melting three-dimensional-printed Ti-6Al-4V. *Scientific Reports*, 6, 26039, 2016. doi: 10.1038/srep26039.

Tao, F., Zhang, L., Venkatesh, V.C., Luo, Y., and Cheng, Y. Cloud manufacturing: A computing and service-oriented manufacturing model. *Proceedings of Institute of Mechanical Engineering Part B: Journal Engineering Manufacture*, 225, no. 4, 1969–1976, 2011.

Telea, A., and Jalba, A. Voxel-based assessment of printability of 3D shapes. In: *10th International Symposium on Mathematical Morphology and Its Applications to Image and Signal Processing*, Verbania-Intra, Italy, July 6–8, 2011.

Tian R., Liu, S. and Zhang, Y. Research on fast grouping slice algorithm for STL model in rapid prototyping. *IOP Conference Series: Journal of Physics: Conference Series* 1074, 012165, 2018. doi:10.1088/1742-6596/1074/1/012165

Todai, M., Nagase, T., Hori, T., Matsugaki, A., Sekita, A., Nakano, T. Novel TiNbTaZrMo High-Entropy Alloys for Metallic Biomaterials. *Scripta Materialia*, 129, 65–68, 2017. doi:10.1016/j.scriptamat.2016.10.028.

Van Arkel, A.C. History and extractive metallurgy. In: *Titanium—Physical Metallurgy, Processing, and Applications*. ASM International: Cleveland, OH, 2015.

Vilaro, T, Colin, C., and Bartout, J.D. As-fabricated and heat-treated microstructures of the Ti-6Al-4V alloy processed by selective laser melting. *Metallurgical and Materials Transactions A*, 42, no.10, 3190–3199, 2011.

Yadroitsev, I., Thivillon, L., Bertrand, P., and Smurov, I. Strategy of manufacturing components with designed internal structure by selective laser melting of metallic powder. *Applied Surface Science*, 254, no. 4, 980–983, 2007.

Yam, R.C.M., Tse, P.W., Li, L., and Tu, P. Intelligent predictive decision support system for condition-based maintenance. *International Journal of Advanced Manufacturing Technology*, 17, 383–391, 2001.

Zhang, L.C., Klemm, D., Eckert, J., Hao, Y.L., and Sercombe, T.B. Manufacture by selective laser melting and mechanical behavior of a biomedical Ti–24Nb–4Zr–8Sn alloy. *Scripta Materialia*, 65, no. 1, 21–24, 2011. ISSN 1359–6462, https://doi.org/10.1016/j.scriptamat.2011.03.024.

Zhao, S., Li, S.J., Hou, W.T., Hao, Y.L., Yang, R., and Misra, R.D.K. The influence of cell morphology on the compressive fatigue behavior of Ti-6Al-4V meshes fabricated by electron beam melting. *Journal of Mechanical Behavior of Bio-Medical Materials*, 59, 251–264, 2016.

Zhou, H., and Wu, J. New optimization method of STL model slice contour on rapid prototyping. *Manufacturing Automation*, 37, no. 12, 25–27, 2015.

Chapter 10

Additive manufacturing of polymer-based functionally graded materials

Mohit Kumar, Neha Choudhary, and Varun Sharma
Indian Institute of Technology (Roorkee)

CONTENTS

10.1 Introduction	167
10.1.1 Homogeneous composition	168
10.1.2 Heterogeneous composition	169
10.2 Additive manufacturing techniques for functionally graded materials object	172
10.2.1 VAT photopolymerization	173
10.2.2 Material extrusion process	177
10.2.3 Powder bed fusion process	179
10.2.4 Material jetting process	179
10.3 Conclusion and future direction	182
References	182

10.1 INTRODUCTION

Functionally graded materials (FGMs) are a class of advanced engineering materials characterized by gradually varying mechanical and microstructural properties specially designed for specific functions [1]. FGMs have been used not only recently but also for thousands of years; for example, bamboo used in construction and decoration. Bamboo is an functionally graded materials (FGM) because fibers' density increases from 10% to 15% at the culm's inner wall to ~60% at the outer wall [2]. In 1972, researchers investigated the potential applications of structural properties of gradient polymeric composite materials [3]. But at that time, the development of these materials was procrastinated due to limited manufacturing techniques. A decade later, in 1984, the scientific term "functionally graded material" was first introduced in Japan to develop and implement thermal barrier materials [4,5]. Afterward, the capacity to generate materials with specialized qualities acceptable for several high-tech applications like automobile, aerospace, nuclear, and biotechnology industries has sparked

DOI: 10.1201/9781003258391-10

168 Additive Manufacturing

The general idea of structural gradients FGM was initially proposed for polymeric materials.	The concept of FGM was first considered in Japan during the design of a space shuttle.	Sendai Group proposed a concept of metallic FGM (Nino, Koizumi and Hirai).	Establishing the concept of FGM.
1972	**1983**	**1984**	**1985**
		Regularly, a conference is held every two years	
1986	**1987**	**1990**	**2021**
Investigation and research conducted for FGM (with Special Coordination Funds for Promoting Science and Technology)'	Launching a National Project called FGM Part I (with Special Coordination Funds for Promoting Science and Technology)	The 1st International Conference on Functionally Graded Materials (FGM 1990) in Sendai, Japan	The 16th International Conference on Functionally Graded Materials (FGM 2021) in Hartford, USA

Figure 10.1 A history of significant events in the research and development of functionally graded materials [5].

renewed interest in FGMs. A detailed timeline overview of the research and development of FGMs is depicted in Figure 10.1.

FGMs can be developed with several conventional methods, but they fail to get low-cost, complex shapes and uniform and precise composition of material gradient over the structure of FGMs. Therefore, a high degree of automation and preparation methods is necessary. The additive manufacturing (AM) processes can fulfill all these requirements. Nowadays, AM has emerged as a novel method for producing intricate and large quantities of FGM parts. The AM system joins materials point by point and layer by layer under the instruction of the planned tool path, eventually producing a 3D part. AM greatly reduces the demand for tooling and assembly procedures. More time and effort can be saved as a result. AM is an environmentally favorable form of production since it produces less waste [6,7].

The advancement of AM technologies allows for the strategic control of the density and directionality of material deposition within a complex 3D distribution. It also produces a seamless monolithic structure by varying deposition density and orientations with the combination of various materials. Three categories can be used to categorize the potential microstructural gradient compositions of FGMs that AM can produce:

(a) homogeneous composition with variable densification,
(b) heterogeneous composition by concurrently combining two or more materials through a gradual transition,
(c) heterogeneous composition using a combination of variable densification.

10.1.1 Homogeneous composition

Homogeneous composition–based FGMs can be created by varying porosity or density gradients with a single material. It is known as varied densification or porosity gradient FGMs. Stiffness and elasticity are influenced

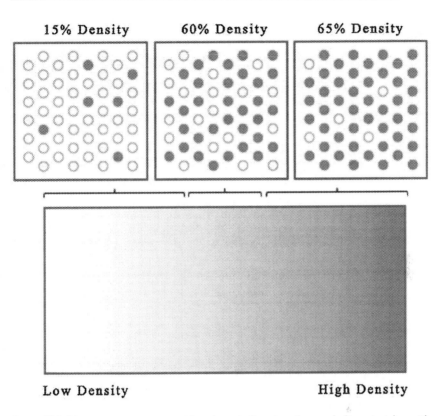

Figure 10.2 Homogeneous composition–based functionally graded materials with variable densification [8].

by the magnitude, directionality, and density concentration of the material substance in a monolithic anisotropic composite structure, as shown in Figure 10.2. Varied densification FGMs can be inspired by biologically mimicking nature's materials, including tissue variation in muscle, spongy trabecular structure in bone, radial density gradients in palm trees, etc. Because of the density gradients (Ex- from a solid exterior to a porous core), these FGMs have a high strength-to-weight ratio, which allows them to be lighter while remaining more robust and efficient.

10.1.2 Heterogeneous composition

In heterogeneous composition–based FGMs, two or more materials can be used to fabricate the object. Furthermore, it can be classified into two categories: multi-material FGMs through a gradual transition (chemical or composition gradient) and multi-material FGMs with variable densification (or microstructure gradient). The overall functions and qualities of FGMs are controlled by the geometric and material arrangement of the phases. Improved interfacial bonds between different or incompatible materials are the goal of AM in multi-material FGMs. Through a heterogeneous compositional transition from a scattered to a linked second phase structure, graded with discrete compositional parameters or smooth concentration gradients, distinct barriers can be eliminated. Thus, it is possible to prevent common failures like delamination and cracks brought on by the surface tension that is experienced by conventional multi-material FGMs as a result of discrete changes in material properties.

The engineering applications that call for smooth graded qualities from the FGMs are shown in Figure 10.3. A tooth and more specifically dental crowns are two good examples of how FGM is applied. High ductility is required on the inner surface in order to prevent wear and brittleness, while high hardness is required on the outer surface in order to prevent fatigue. While the concept of FGM was initially intended for materials that could withstand heat, these materials have since been used to control pressure, deformation, corrosion and wear as well as to lessen the concentration of stress through the use of a smooth transition graded across the whole product dimensions. By supporting research in a range of initiatives, many engineering sectors, including those in the automotive, biomedical, and aviation industries, have greatly aided in the development of FGMs via AM.

The printed component can have the best qualities of both materials by three-dimensionally fusing one material to another using a dynamic gradient. It can have a transitional weight while still maintaining its mechanical or

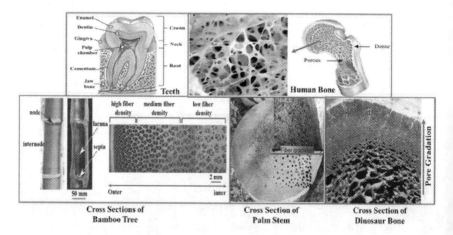

Figure 10.3 Examples of naturally occurring functionally graded materials that have structures with graded qualities [5].

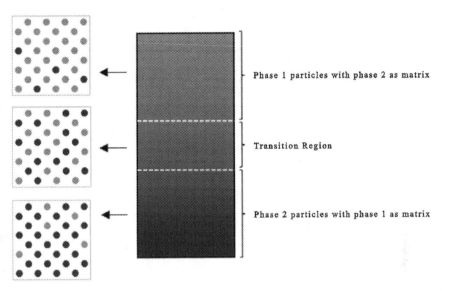

Figure 10.4 Continuously graded microstructure between two materials in a multi-material functionally graded material [8].

physical attributes including wear resistance, toughness, impact resistance, and physical, chemical, or biochemical capabilities. It is no longer necessary for heterogeneous combinations of materials to rely on their intrinsic qualities to produce the component's desired properties.

Figure 10.4 shows how the materials in the multi-material FGM transition smoothly and seamlessly from 0% at one end to 100% at the other. Monitoring the ratios of two or more materials that are mixed during deposition and before curing allows for continuous variation within the 3D space. The compositional variation, on the other hand, must be controlled by the computer program.

There are four different ways to design heterogeneous compositional gradients: a transition between two materials, three materials or more, switching composition between different locations, or combining density and compositional gradation, as depicted in Figure 10.5.

The size of the gradient vector, the geometry, and the distribution of the equipotential surfaces are some of the primary design factors for FGM. The gradient's direction within the composition further influences the component's characteristics and functioning. According to 1D, 2D, and 3D, as shown in Figure 10.6, the volumetric gradient's design and types may be categorized, and the materials can be distributed either uniformly or according to certain patterns.

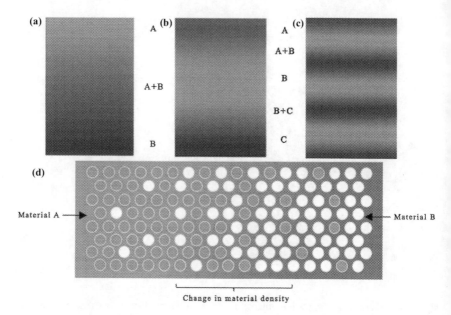

Figure 10.5 (a) Multi-material functionally graded material (FGM) (two materials), (b) multi-material FGM (three materials), (c) switched compositions, and (d) a heterogeneous material with a combination of density and compositional gradation [8].

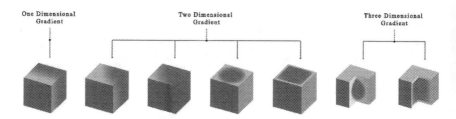

Figure 10.6 Gradient classification types [8].

10.2 ADDITIVE MANUFACTURING TECHNIQUES FOR FUNCTIONALLY GRADED MATERIALS OBJECT

The term additive manufacturing describes a method of producing products in which materials are added in a regulated manner to produce the required shape and size. This method allows for the direct production of near-net-shape products without the use of molds, jigs, or fixtures as in traditional manufacturing. Furthermore, by allowing for design flexibility, AM outperforms traditional manufacturing techniques in the preparation of geometrically complex products at a lower cost and in less time. Because of the rapid advancement of AM techniques, they are no longer limited to single-phase materials; FGM can be used, giving rise to the terminology, Functionally Graded Additive Manufacturing (FGAM). To fulfill an intended function, it entails gradationally changing the material arrangement within a component. AM has made it possible to supply desired material gradations more accurately. Furthermore, AM methods frequently rely on product design and technology to achieve higher performance and better quality. Many AM techniques have lately become widespread in a number of applications needing graded features, such as medical, space, automobile, aerospace, electronics, sports, and architecture applications, as a result of the major growth of FGMs [9].

The following sections discuss and explain the most commonly used AM production technologies for polymeric FGMs.

10.2.1 VAT photopolymerization

VAT photopolymerization (including stereolithography (SLA) and digital light processing (DLP)) layers a photosensitive resin into a 3D solid part using a high-power laser, which converts monomers into cross-linked polymers via the polymerization process. The input material is kept in a vat as a liquid and a computer interface controls the movement of the platform and laser during fabrication. It has the appealing characteristics of producing objects with a high-quality surface finish, dimensional accuracy, and a range of material options. The material distribution in this process along a layer is homogenous. On the other hand, it has the potential to produce functionally graded material. SLA with multi-material object printing is depicted schematically in Figure 10.7.

In order to create multi-material FGMs, however, the setup must be modified to allow for the deposition of many materials simultaneously. New studies

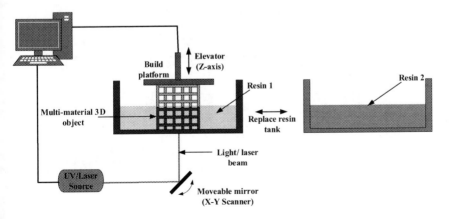

Figure 10.7 Multi-material stereolithography process.

174 Additive Manufacturing

investigating multi-vat systems can be found nowadays. In this context, a new SLA setup was created that includes multiple vats with different resins equipped with a rotating system about the vertical axis and an automatic leveling system to produce multi-material complex parts (refer Figure 10.8a) [10]. The rook or castle was used for multi-materials' (three resins) fabrication (refer Figure 10.8b). The three resins are the subparts and were constructed on platform or directly on previously printed layers to produce the whole part. Each subpart was separately sliced and the sequence of printing for different

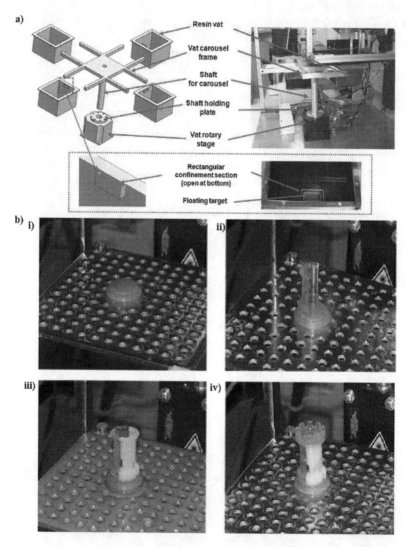

Figure 10.8 (a) Rotating vat setup and (b) fabricated functionally graded material part: (i) Model 1 with ProtoTherm™12120, (ii) Model 2 using WaterShed™ 11120, (iii) Model 3 by Somos14120, and (iv) Model 4 with ProtoTherm™12120 [10].

material was planned. The total time of printing of the model is ~24 hours. The developed system has the application of producing multi-material parts such as surgical models, electronic devices, and tissue engineering scaffolds.

SLA was used to create a Si_3N_4-Al_2O_3-based functionally graded ceramic composition. The setup used for printing consists of an SL-3D printer, different paste barrels, and a manipulator. The manipulator system was used to transfer the gradient pastes to the suspension. The different volume percentages of ceramics were used to prepare the paste, and the process parameters were optimized based on the rheology of paste and ultraviolet (UV) curing factors. Finally, FGM part was printed successfully, as shown in Figure 10.9 [11]. Another work involved curing two distinct resins (one for photochemical polymerization activation and the other for reaction inhabitation) at two different wavelengths to create volumetric patterns. It was found that grading in part properties in a single phase of manufacturing may be possible due to regulated, concurrent photoinitiation, and photoinhibition [12]. Stereolithography was successfully integrated with topology optimization using unit cell design to create FGMs with regulated stiffness [13]. Eight functionally graded unit cells were printed (refer Figure 10.10), and compression test was performed. It was found that depending on the type and arrangement of unit cells, the compression performance of designed functionally graded materials exhibits distinct behavior [13]. With the use of grayscale masked SLA, altering the degree of curing and light intensity grading in some properties such as porosity, density, and stiffness can be achieved [14].

A unique DLP setup was created by using a moveable glass plate to allow for the deposition, followed by the air-jet cleaning procedure to reduce contamination and slurry waste in order to print two distinct materials [15]. Similarly, the fabrication of FGMs from resin-containing magnetite (Fe_3O_4) magnetic particles involved a DLP technique (refer Figure 10.11). In order to obtain varying degrees of the gradient in magnetic particles, the distance between the magnet and the print bed was varied. This technique enables the production of FGMs without the need to modify a single material DLP machine in an expensive way [16].

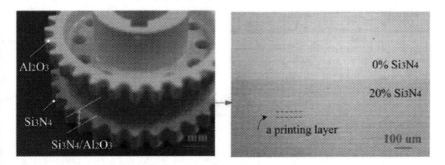

Figure 10.9 Functionally graded part via SLA-3D technology [11].

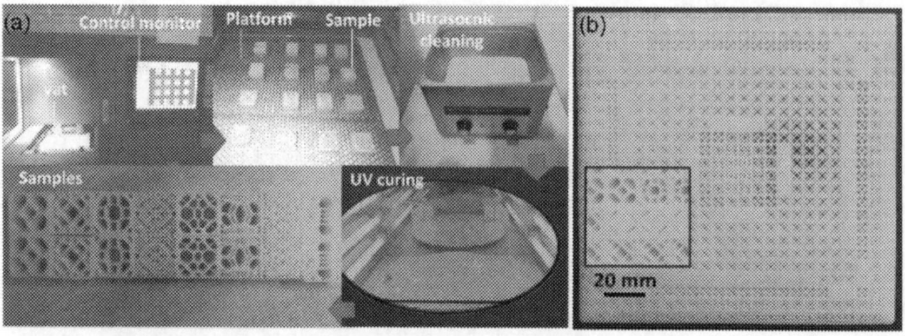

176   Additive Manufacturing

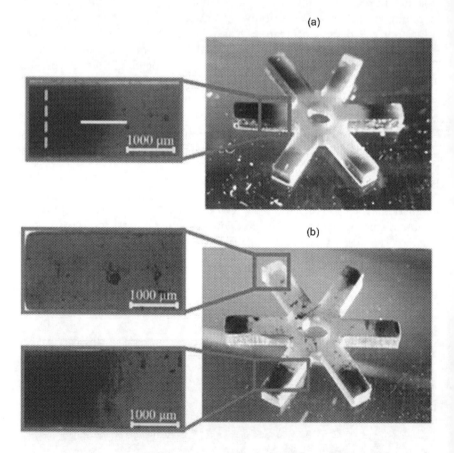

Figure 10.10 (a) Production and post-processing of functionally graded material samples and (b) selective laser sintering printed FG structure with varying stiffness [13].

Figure 10.11 Printed structure with graded magnetic particles (a) with tubular magnet and (b) with circular array [16].

10.2.2 Material extrusion process

The multi-material fused deposition modeling (FDM) process illustrated in Figure 10.12 works on the following principles: a thermoplastic filament material is melted and selectively deposited by a computer control nozzle, forming layers with a pre-designed cross section and cools rapidly. When a layer is fully applied, the printing device will need to shift the component upward or downward in order to lay another layer, completing one layer at a time until the entire part is manufactured.

FDM machines with multiple nozzles have the potential to additively fabricate functionally graded material objects. Researchers constructed an FGM utilizing two distinct resins composed of differing proportions of methacrylates and acrylates dispensed from two different dispensers [17]. In line with this, a triple-extruder setup was developed to prepare FGMs produced using a unique extrusion technique called the freeze-form extrusion fabrication technique, as shown in Figure 10.13a. The efficiency of the developed system was first shown by fabricating parts made of limestone ($CaCO_3$) with graduated shades. Next, the pink and green graded part with Al_2O_3 and ZrO_2 ceramics was created (refer Figure 10.13b) by varying the relative flow rates of the corresponding plungers. The created part underwent post-processing, and material compositions were verified via energy dispersive spectroscopy (EDS) analysis [18]. Similarly, SiC/Al_2O_3 reinforced aluminium (Al)-based FGM products were fabricated using FDM and investment casting to evaluate the mechanical properties of the resulting graded structure [19]. The functionally graded FDM components can be modeled and simulated to analyze the effect of process parameters under various load scenarios [20].

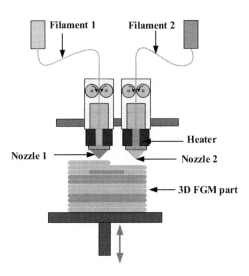

Figure 10.12 Schematic diagram of multi-material fused deposition modeling process.

178 Additive Manufacturing

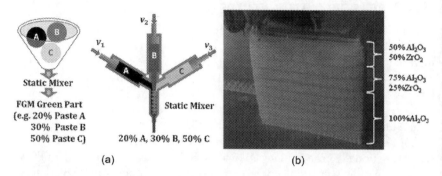

Figure 10.13 (a) Triple-extruder design and (b) fabricated functionally graded material test bar [18].

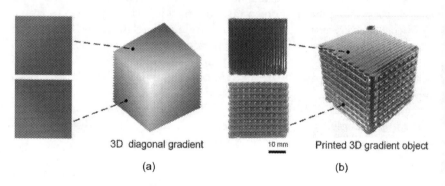

Figure 10.14 3D-printed object: (a) gray-scale and (b) gradient color representation [23].

The advancement of contemporary tools and technologies helped to create nonlinear gradient materials. In this context, an artificial tendon-muscle-tendon system with regionally linear changing colors was successfully created using curable acrylate urethanes and hydrogel [21]. Similarly, researchers demonstrated the feasibility of constructing heterogeneous tissue in medical applications using the FDM process with four different nozzles. The hydrogel material allows for the deposition of functional gradient scaffolds in particular [22]. In another work, mathematical functions were used to illustrate the graded material qualities during printing. Then, nano-sized Al_2O_3 particles were supplied to the printer digitally to create one-dimension, two-dimension, and three-dimension graded objects in accordance with controlling code creation and gray-scale representation (refer Figure 10.14) [23].

10.2.3 Powder bed fusion process

Selective laser sintering (SLS) is able to generate complex components with variable mechanical properties. SLS selectively scans the powders along a predetermined path and melts and sinters the scanned powders. Then, a layer of powders with a thickness of microns is delivered, and the laser sinters it again. This process takes place until the whole part gets completed. SLS usually manufactures FGMs with a variation in composition/constituent along the perpendicular direction to layers, as shown in Figure 10.15.

The Functionally graded (FG) polymer composites, Nylon-11 with different volume ratios (0%–30%), of glass beads were fabricated for 1D FGM, as shown in Figure 10.16. The design of experiments was used to investigate the influence of SLS process factors on the tensile and compressive modulus. The result showed that the addition of fumed silica enhanced the mechanical property of FGMs, making them stiff but brittle [24]. Similarly, nylon-11 with silica particles with volume ratio variation of 0%–10% was also created and studied for mechanical properties [25].

Another work involved creating functionally graded scaffolds with various structural arrangements and porosities for tissue engineering using PCL as the base material. It was found that the stiffness of the scaffold is similar to that of the cancellous bone of the maxillofacial region. The cytotoxicity test revealed that the SLS process was able to create scaffolds without contaminating the nonproprietary PCL material [26].

10.2.4 Material jetting process

The material jetting technique creates parts similar to a 2D inkjet printer. A variety of printheads are used in polyjet printing, and they move in both X and Y directions while printing. The print heads spray photopolymerizable

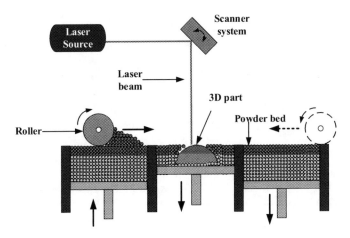

Figure 10.15 Schematic diagram of multi-material selective laser sintering.

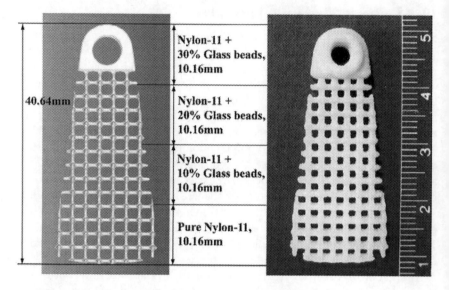

Figure 10.16 A functionally graded–based rotator cuff scaffold printed through selective laser sintering [24].

material onto the bed on demand, the roller smoothes the surface of the sprayed materials, and the UV light cures the material. Following the completion of one layer, the table accurately drops by layer height, and print heads continue to spray the photosensitive polymer material for the next layer's printing. The steps are repeated until the entire part has been constructed. A multi-head setup with different materials is required for FGM printing, as shown schematically in Figure 10.17.

FGMs can be printed with different graded qualities, including color, transparency, and stiffness, through the material jetting process. In order to create the color-graded FGM yellow-magenta, a transparent material named mix Tango Plus, along with Vero yellow and Vero magenta, two opaque materials, has been used. It was found that the fading of color saturation and opacity causes the gradation's density to decrease [27]. Similarly, circular and rectangular graded zones were created by using Tango Black+ material and Vero White and Acrylonitrile Butadiene Styrene (ABS)-based material. It showed that there were only a few variations between the strain patterns predicted by Finite element analysis (FEA) and those discovered experimentally in the tensile test [28]. Another work has been done showing the material gradient pattern's (stepwise and continuous) effect on the fatigue strength of FGMs, as shown in Figure 10.18. It was observed that the stepwise gradient showed better fatigue life and poor properties at the interface region [29]. Researchers have constructed meso (1 mm) digital

Additive manufacturing of graded materials   181

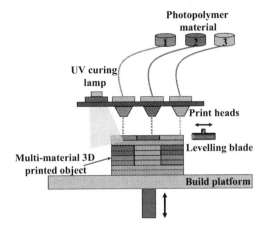

Figure 10.17 Schematic representation of multi-material jetting technique.

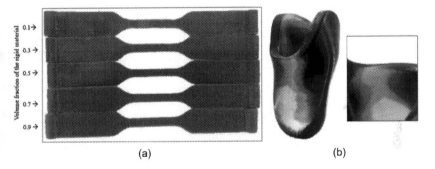

Figure 10.18 (a) Graded fatigue specimens with volume fraction [29] and (b) fabricated socket showing the material transition [31].

material through material jetting by controlling the fashion of polymer deposition at the voxel scale [30]. Similarly, a bit map printing technique through material jetting was used to fabricate a prosthetic socket, as shown in Figure 10.18b. Pressure-sensing components are also created at the same time as the socket, offering opportunities for assessing how well they fit. The printer's native resolution is used to set voxel resolution in this work, and this removes the need for slicing and path planning [31]. As the type and amount of materials sprayed by print heads can be easily and flexibly adjusted, complex FGMs with gradient compositions in multiple directions can be fabricated. To better meet the fabrication needs of FGM parts, the material database should be expanded.

10.3 CONCLUSION AND FUTURE DIRECTION

FGMs have emerged as a mainstay of contemporary materials research, finding wide-ranging uses in the aerospace, defense, energy, and medical industries. This comprehensive review of the literature on FGMs based on production techniques sheds some light on the variables that ultimately determine their characteristics. As a result, there is less information available about polymer FGM-processing techniques, applications, and qualities as compared to ceramic or metal-based systems. The development of new families of composites with customized properties across the thickness and the expansion of the technical uses of polymeric materials are both possible to gradient microstructure in nano-reinforced polymer composites. With cutting-edge AM methods and incorporating nanomaterials like graphene, carbon nanotubes, and boron nitride, biopolymeric-based functionally graded structures can be created that have improved mechanical, physical, and chemical properties for use in biological applications.

Other underutilized techniques have the potential to be used to create the next generation of gradient structures, in addition to the AM processes and tried-and-true approaches for fabricating FGMs. There are many opportunities to produce materials with novel gradient compositions for applications where the multifunctional features of such gradients are most advantageous to the significant advancements in AM techniques.

The scientific and technological developments offer hope for future work as follows:

- The advancement of artificial intelligence techniques, which may aid in the creation of materials and processes that are optimal and may result in in-situ microstructural and physical property optimization.
- The creation and application of novel diagnostic methods, which, when appropriately integrated with feedback control loops, can be utilized to keep process conditions constant.
- Thermodynamic databases, such as Calculation of Phase Diagrams (CALPHAD), to create alloy compositions in advance can be utilized in AM to predict phase composition and stability.
- The design and optimization of AM techniques can be used to produce multidimensional, intricate, and multifunctional FGMs that can be used in harsh working environments.

REFERENCES

1. Oxman N, Keating S, Tsai E (2011) Functionally Graded Rapid Prototyping. In: *Innovative Developments in Virtual and Physical Prototyping.* CRC Press, pp 483–489.

2. Gottron J, Harries KA, Xu Q (2014) Creep behaviour of bamboo. *Construction and Building Materials* 66:79–88. https://doi.org/10.1016/j. conbuildmat.2014.05.024.

3. Shen M, Bever MB (1972) Gradients in polymeric materials. *Journal of Materials Science* 7:741–746. https://doi.org/10.1007/BF00549902.

4. Koizumi M (1997) FGM activities in Japan. *Composites Part B: Engineering* 28:1–4. https://doi.org/10.1016/S1359–8368(96)00016–9.

5. Saleh B, Jiang J, Fathi R, Al-hababi T, Xu Q, Wang L, Song D, Ma A (2020) 30 years of functionally graded materials: An overview of manufacturing methods. *Applications and Future Challenges. Composites Part B: Engineering* 201:108376. https://doi.org/10.1016/j.compositesb.2020.108376.

6. Kumar M, Sharma V (2021) Additive manufacturing techniques for the fabrication of tissue engineering scaffolds: A review. *Rapid Prototyping Journal* 27:1230–1272. https://doi.org/10.1108/RPJ-01-2021–0011.

7. Choudhary N, Sharma V, Kumar P (2021) Reinforcement of polylactic acid with bioceramics (alumina and YSZ composites) and their thermomechanical and physical properties for biomedical application. *Journal of Vinyl and Additive Technology* 27:612–625. https://doi.org/10.1002/vnl.21837.

8. Loh GH, Pei E, Harrison D, Monzón MD (2018) An overview of functionally graded additive manufacturing. *Additive Manufacturing* 23:34–44. https:// doi.org/10.1016/j.addma.2018.06.023.

9. Yan L, Chen Y, Liou F (2020) Additive manufacturing of functionally graded metallic materials using laser metal deposition. *Additive Manufacturing* 31:100901. https://doi.org/10.1016/j.addma.2019.100901.

10. Choi J-W, Kim H-C, Wicker R (2011) Multi-material stereolithography. *Journal of Materials Processing Technology* 211:318–328. https://doi. org/10.1016/j.jmatprotec.2010.10.003.

11. Xing H, Zou B, Liu X, Wang X, Huang C, Hu Y (2020) Fabrication strategy of complicated Al2O3-Si3N4 functionally graded materials by stereolithography 3D printing. *Journal of the European Ceramic Society* 40:5797–5809. https://doi.org/10.1016/j.jeurceramsoc.2020.05.022.

12. De Beer MP, Van Der Laan HL, Cole MA, Whelan RJ, Burns MA, Scott TF (2019) Rapid, continuous additive manufacturing by volumetric polymerization inhibition patterning. *Science Advances* 5:1–9. https://doi.org/10.1126/ sciadv.aau8723.

13. Liu T, Guessasma S, Zhu J, Zhang W, Belhabib S (2018) Functionally graded materials from topology optimisation and stereolithography. *European Polymer Journal* 108:199–211. https://doi.org/10.1016/j. eurpolymj.2018.08.038.

14. Valizadeh I, Al Aboud A, Dörsam E, Weeger O (2021) Tailoring of functionally graded hyperelastic materials via grayscale mask stereolithography 3D printing. *Additive Manufacturing* 47:102108. https://doi.org/10.1016/j. addma.2021.102108.

15. Kowsari K, Akbari S, Wang D, Fang NX, Ge Q (2018) High-efficiency high-resolution multimaterial fabrication for digital light processing-based three-dimensional printing. *3D Printing and Additive Manufacturing* 5:185–193. https://doi.org/10.1089/3dp.2018.0004.

184 Additive Manufacturing

16. Safaee S, Chen R (2019) Investigation of a magnetic field-assisted digital-light-processing stereolithography for functionally graded materials. *Procedia Manufacturing* 34:731–737. https://doi.org/10.1016/j.promfg.2019.06.229.

17. Kokkinis D, Bouville F, Studart AR (2018) 3D printing of materials with tunable failure via bioinspired mechanical gradients. *Advanced Materials* 30:1705808. https://doi.org/10.1002/adma.201705808.

18. Leu MC, Deuser BK, Tang L, Landers RG, Hilmas GE, Watts JL (2012) Freeze-form extrusion fabrication of functionally graded materials. *CIRP Annals* 61:223–226. https://doi.org/10.1016/j.cirp.2012.03.050.

19. Singh N, Singh R, Ahuja IPS (2018) On development of functionally graded material through fused deposition modelling assisted investment casting from Al2O3/SiC reinforced waste low density polyethylene. *Transactions of the Indian Institute of Metals* 71:2479–2485. https://doi.org/10.1007/s12666-018-1378-9.

20. Srivastava M, Maheshwari S, Kundra TK, Rathee S, Yashaswi R, Kumar Sharma S (2016) Virtual design, modelling and analysis of functionally graded materials by fused deposition modeling. *Materials Today: Proceedings* 3:3660–3665. https://doi.org/10.1016/j.matpr.2016.11.010.

21. Bakarich SE, Gorkin R, Gately R, Naficy S, in het Panhuis M, Spinks GM (2017) 3D printing of tough hydrogel composites with spatially varying materials properties. *Additive Manufacturing* 14:24–30. https://doi.org/10.1016/j.addma.2016.12.003.

22. Khalil S, Nam J, Sun W (2005) Multi-nozzle deposition for construction of 3D biopolymer tissue scaffolds. *Rapid Prototyping Journal* 11:9–17. https://doi.org/10.1108/13552540510573347.

23. Ren L, Song Z, Liu H, Han Q, Zhao C, Derby B, Liu Q, Ren L (2018) 3D printing of materials with spatially non-linearly varying properties. *Materials & Design* 156:470–479. https://doi.org/10.1016/j.matdes.2018.07.012.

24. Chung H, Das S (2006) Processing and properties of glass bead particulate-filled functionally graded Nylon-11 composites produced by selective laser sintering. *Materials Science and Engineering: A* 437:226–234. https://doi.org/10.1016/j.msea.2006.07.112.

25. Chung H, Das S (2008) Functionally graded Nylon-11/silica nanocomposites produced by selective laser sintering. *Materials Science and Engineering: A* 487:251–257. https://doi.org/10.1016/j.msea.2007.10.082.

26. Sudarmadji N, Tan JY, Leong KF, Chua CK, Loh YT (2011) Investigation of the mechanical properties and porosity relationships in selective laser-sintered polyhedral for functionally graded scaffolds. *Acta Biomaterialia* 7:530–537. https://doi.org/10.1016/j.actbio.2010.09.024.

27. Loh GH, Pei E, Harrison D, Monzón MD (2018) An overview of functionally graded additive manufacturing. *Additive Manufacturing* 23:34–44. https://doi.org/10.1016/j.addma.2018.06.023.

28. Salcedo E, Baek D, Berndt A, Ryu JE (2018) Simulation and validation of three dimension functionally graded materials by material jetting. *Additive Manufacturing* 22:351–359. https://doi.org/10.1016/j.addma.2018.05.027.

29. Kaweesa DV, Meisel NA (2018) Quantifying fatigue property changes in material jetted parts due to functionally graded material interface design. *Additive Manufacturing* 21:141–149. https://doi.org/10.1016/j.addma.2018.03.011.

30. Ituarte IF, Boddeti N, Hassani V, Dunn ML, Rosen DW (2019) Design and additive manufacture of functionally graded structures based on digital materials. *Additive Manufacturing* 30:100839. https://doi.org/10.1016/j.addma.2019.100839.
31. Doubrovski EL, Tsai EY, Dikovsky D, Geraedts JMP, Herr H, Oxman N (2015) Voxel-based fabrication through material property mapping: A design method for bitmap printing. *Computer-Aided Design* 60:3–13. https://doi.org/10.1016/j.cad.2014.05.010.

Chapter 11

Processing techniques, principles, and applications of additive manufacturing

Sumit Kumar Sharma and Ranjan Mandal
BIT Sindri

Amarish Kumar Shukla
Indian Institute of Technology (Kharagpur)

CONTENTS

List of abbreviations	188
11.1 Introduction	188
11.1.1 Principle of additive manufacturing	189
11.1.2 Advantages and limitations of additive manufacturing	189
11.1.3 Additive manufacturing development from novelty to mainstream manufacturing	190
11.2 Classification of additive manufacturing	191
11.2.1 Directed energy deposition (DED)	191
11.2.2 Material jetting	192
11.2.3 Material extrusion	192
11.2.4 Powder bed fusion	192
11.2.4.1 Direct metal laser sintering (DMLS)	193
11.2.4.2 Electron beam melting (EBM)	194
11.2.5 Powder jetting	195
11.2.6 Vat polymerization	196
11.2.7 Sheet lamination	196
11.3 Application of additive manufacturing	197
11.3.1 Medical	197
11.3.2 Energy	197
11.3.3 Transportation	197
11.3.3.1 Automotive sector	197
11.3.3.2 Aerospace/aviation sector	197
11.4 Development of additive manufacturing	199
11.4.1 Build process	199
11.4.2 Part validation	199
11.4.3 Conduct virtual prototype testing	199
11.5 Summary and future work of additive manufacturing	200
11.6 Funding declaration	200
References	200

DOI: 10.1201/9781003258391-11

LIST OF ABBREVIATIONS

3D	Three-dimensional
AM	Additive manufacturing
BJ	Binder jetting
CAD	Computer aided design
DED	Directed energy deposition
DMLS	Direct metal laser sintering
EBW	Electron beam welding
PBF	Powder bed fusion
SLM	Selective laser melting
UTS	Ultimate tensile strength
YS	Yield strength

11.1 INTRODUCTION

Additive manufacturing is getting attention not only in the manufacturing industries but also in various sectors such as the medical, transportation, energy, etc. due to its better productivity, functionality, and better potential. Additive manufacturing is reliable, sustainable, and eco-friendly (Colorado et al., 2020; Ian GibsonDavid W. RosenBrent Stucker, 2010; Vafadar et al., 2021). Additive manufacturing, commonly known as 3D printing refers to a technique capable to develop layer-by-layer building geometry in a single step (Ma et al., 2013; SONG et al., 2011; Xie et al., 2013). Since the development of the additive manufacturing set up three decades ago, several processing methods have advanced significantly, and studying the properties of developed additive manufactured products. Hence, the microstructure and properties of the building part have been understood in detail, it became easier to optimize the processing parameters.

(Peng et al., 2018)

Several materials, including titanium alloys, nickel alloys, aluminium alloys, steel, etc., have used additive manufacturing (AM) extensively in the manufacturing industries. Several materials and alloys have been processed by AM. However, various alloys and metals are still processed by conventional processes such as casting, forming, machining, etc. (Gibson et al., 2010; Hopkinson et al., 2006). AM also provides freedom from the design constraint in the manufacturing industry and develops part of the complicated design with reduced weight and high material yield (Huang et al., 2016). Because AM techniques can utilize recycled feedstock materials, they have the potential to achieve zero waste. As a result, emissions are reduced because fewer raw materials are required. Furthermore, unlike traditional manufacturing technologies that require coolants or lubricants, AM techniques do not necessitate the direct use of harmful chemicals (Colorado et al., 2020).

11.1.1 Principle of additive manufacturing

AM covers all the phases of all the techniques that develop layer-by-layer material at the small level to get the desired object without using the traditional subtractive method of metal removal. During the research stage, the American Society for Testing and Materials (ASTM) committee on additive layer technologies standardized AM. Models are created with AM Technologies by fusing and sintering materials in specified layers without the use of tools. AM allows for the creation of complicated geometries, including internal part detail, and it cannot be achieved by using conventional methods. Slicing computer aided design (CAD) data with professional software creates these AM models (Olsén et al., 2018; Zhong et al., 2017). The principles of all AM systems are the same. The thickness of the layer varies depending on the specifications and machine employed, and can range from 10 to 200 m. Figure 11.1 shows the flowchart for obtaining components by AM process.

1) Basically it's creating a three-dimensional (3D) model by using a computer aided design (CAD).
2) Hence, slice the solid part into a two-dimensional (2D) planer layer.
3) It sends printing order to the 3D printer.

11.1.2 Advantages and limitations of additive manufacturing

AM has various advantages over other traditional techniques, but it also has certain limitations as well. Table 11.1 lists the benefits and drawbacks of the AM approach in comparison to other conventional techniques. As a result, before AM can be employed in mass production, it must improve in terms of overall efficiency and process parameters control (Gu et al., 2012).

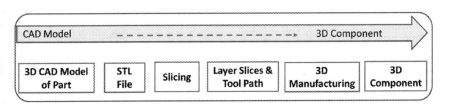

Figure 11.1 Flowchart for obtaining components by additive manufacturing process (Tolosa et al., 2010; Yusuf et al., 2017; Campbell & Ivanova, 2013).

190 Additive Manufacturing

Table 11.1 List of advantages and limitations of additive manufacturing processes over conventional process

Advantages	Limitations
• Fast production rate	• Size limitation
• Free form assembly	• Time consumption
• Complex design freedom	• Surface roughness
• Reliability and repeatability	• Capital cost
• Minimum cost	• Power consumption
• Flexibility	
• Sustainability	

The size of the product is one of the limitations of AM. The layer-by-layer manufacture of the specimen requires more time due to the huge size of the sample. Furthermore, the surface morphology of the additively manufactured specimen is rough and ridged. As a result, it proceeds to the post-processing stage. The setup for AM is thought to be expensive in terms of investment. Constant power supply is necessary to create the object; however, power consumption is a major constraint.

11.1.3 Additive manufacturing development from novelty to mainstream manufacturing

With the introduction of polymer 3D printing in the 1990s, the method began to evolve from its origins. The novel material—a plastic filament extruded through a gantry-mounted nozzle—was the system's defining feature. With the introduction of polymer 3D printing and its more durable material, printed items that might potentially withstand functional testing were available. Print resolutions were initially relatively coarse, but as time went on, newer machines began to generate parts that were suitable for use in some applications. A wide variety of metals can now be used in 3D printing thanks to recent advancements in material selections, opening up a variety of production uses. AM has revolutionized prototyping and new product development over the last decade. Three-dimensional printing allows functional items that previously required complicated mould tools to progress from design to reality. Manufacturers can reduce lead times by weeks, if not months, thanks to this technology. Physical tooling, including mould tools, is now made using additive processes, allowing customers to bring their goods to market at previously unheard-of speeds.

Three-dimensional printing is fast gaining traction in the manufacturing industry. Its applications appear to have no bounds, as they are used across a wide range of industries. Personalized healthcare is becoming realized in the medical device manufacturing business as 3D printing is utilized to make medical implants that are created and printed exactly to fit individual patients. Complex jet engine systems have been replaced with single 3D printed parts in the aerospace industry, saving weight and cutting assembly time while enhancing strength. Since its inception, 3D printing has made enormous strides in terms of innovation. Its rise from a revolutionary technique to an industrial mainstay has been nothing short of spectacular. And, if the past two decades are any indication, the manufacturing industry's meteoric rise is far from over.

11.2 CLASSIFICATION OF ADDITIVE MANUFACTURING

In general, AM can be divided into a few different processes. Figure 11.2 shows the categorization of AM processes.

11.2.1 Directed energy deposition (DED)

One of the difficult types of AM to master is this one. The stationary object will be surrounded by a four- or five-axis robotic arm that will deposit molten material. The material has been melted by using laser or electron beam, and it became solidified. Although ceramics and polymers can also be employed, metal powder or wires are the most frequently used material with DED. DED's ability to control grain structure allows us to work with

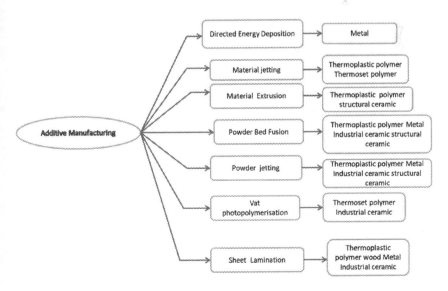

Figure 11.2 Additive manufacturing processes are categorized according to their unique material handling capacities (Manfredi et al., 2014; Michelle, 2021; Nastac et al., 2017).

high level of precision. The surface finish of the product depends on the material opted. Although a powder provides a far better finish than wire in the case of metal, we can get our desired appearance with wire by post-processing. In order to repair or produce parts, Direct Energy Disposition is frequently used.

11.2.2 Material jetting

It's a sort of AM technique in which the material is deposited onto the surface in the form of droplets with the printhead elevated above the platform. After that, the droplets harden, forming a coating. This is done over and over again, layer by layer. Using the drop on demand (DOD) approach, the droplets can be dispensed continuously or individually. Various materials, as well as polymers and waxes, can be used in materials jetting. This method is exact, and it allows us to work with numerous materials on a single project. Material jetting is frequently used to make realistic prototypes or models.

11.2.3 Material extrusion

The material is drawn through a nozzle, heated, and then deposited in a continuous stream in this process, which is commonly employed in low-cost at-home 3D printers. The platform travels up, down, and vertically, while the nozzle moves horizontally. The layers are built in this manner. Temperature and chemical agents can also be used to influence layer bonding. Although material extrusion is frequently seen in low-cost models, it does have potential. Plastics and polymers can be employed to provide robust structural support. This has some drawbacks.

a) Because of the nozzle thickness, accuracy is diminished.
b) Material extrusion is also one of the more time-consuming methods of 3D.

Material jetting is utilized by many automobile firms to develop manufacturing equipment for assembly lines.

11.2.4 Powder bed fusion

Powder bed fusion (PBF) is a type of AM in which powder fusion takes place on the stage. A layer of a certain powder is spread across the stage. The powder is fused using a high-intensity thermal source, such as an electron beam or laser, before a second layer of thin coating is applied with a roller blade. The unfused or unmelted powder is removed after the required shape is completed and reused in the next fabrication process. This PBF technology was used to print the wing brackets for the aero plane. This layering process is then repeated.

These have some variations:

a) Direct metal laser sintering (DMLS)
b) Electron beam melting (EBM)

Metals and polymer materials in the form of powder can be employed as support structures, making it a good choice for prototypes. This AM technology is still utilized to make parts for aviation, automobile, etc.

11.2.4.1 Direct metal laser sintering (DMLS)

Another type of PBF process is DMLS, which involves completely melting the metal powder. The printing was done in the same way as the PBF technique, layer by layer, generating thin layers of melted metal powders such as Ti, Cr, Co, Al, and different alloys (Tolosa et al., 2010; Sharma et al., 2020c). After printing, the superfluous powder is removed, leaving a smooth surface with a dense and uniform volume of structural part that requires little post-processing (Gale & Achuhan, 2017). The entire melting of each powder bed results in densities of roughly 100%. Direct Metal Laser Sintering (DMLS) is a material used to make jet engines and other aircraft components (Jiao et al., 2018; Zhong et al., 2017). General Electric (GE) is also using DMLM technology to print the fuel nozzles for its Leading Edge Aviation Propulsion (LEAP) engines, which reduced the number of parts from 20 to just one (Manfredi et al., 2014; Vayre et al., 2012). The DMLM process is depicted schematically in Figure 11.3.

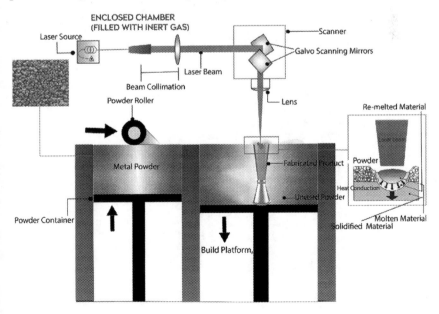

Figure 11.3 Schematic diagram of direct metal laser sintering (Olsén et al., 2018; Tan et al., 2018).

11.2.4.2 Electron beam melting (EBM)

The powder fusion in the EBM process is accomplished using a high-power density electron beam source in a vacuum chamber. The metal powder is melted by a beam of concentrically focused electrons. This method favours the creation of structures with minimum residual stress. The vacuum procedure creates a completely clean and reaction-free environment. EBM is used to produce extremely precise components for the aerospace, automotive, and medical industries (Sharma et al., 2020a, b; Zhong et al., 2017). The EBM technique is used to create biomedical implants. Jet engines and rocket engines both employ EBM equipment. EBM is a more rapid AM method

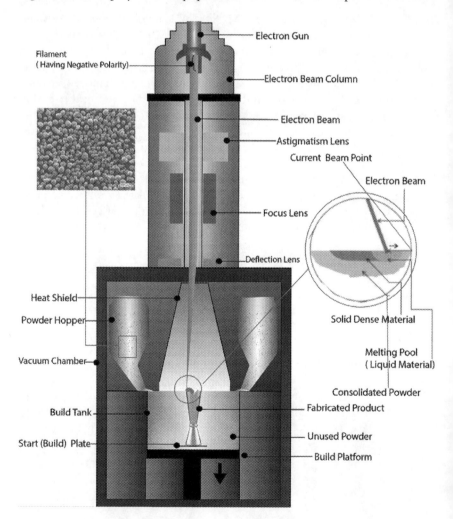

Figure 11.4 Illustration of the layout of electron beam melting (Olsén et al., 2018; Wroe, 2015).

than conventional PBF methods in general, although the deposited layers are thicker and the surface quality is poor. EBM works as an AM technique with a variety of metals, including steel, titanium, copper, cobalt, and carbon (Gu, 2015). Figure 11.4 portrays the EBM process schematics (Olsén et al., 2018).

11.2.5 Powder jetting

This AM uses a binder and a powder-based material. The binder holds the layers together and is often in the liquid form. This is done again and again until the product is complete. We can employ a variety of materials in this procedure, including polymers, ceramics, and metal. It's the quickest AM approach (Gibson et al., 2021). We can adjust the binder powder ratio, for example, if we need a specific quality of material. Figure 11.5 shows the formation of voxel in the binder jetting process. It's possible to manufacture using some materials:

- Gypsum powder (chalk)
- Plastic
- Ceramics
- Metal
- Glass

It's relatively low cost compared to other process.

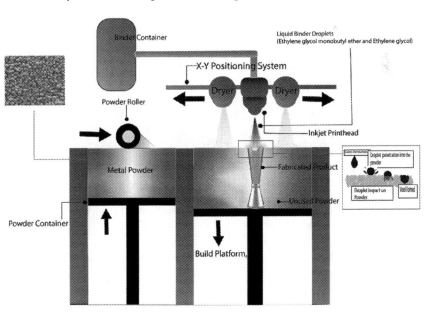

Figure 11.5 The formation of voxels in binder jetting process (Nastac et al., 2017; Olsén et al., 2018; Oropeza & Hart, 2021).

11.2.6 Vat polymerization

It's also known as stereolithography. It's a type of AM process. A vat of liquid photopolymer resin is used in this type of AM. A construction platform moves downwards from the top of the resin, and a laser beam creates a shape in the resin, forming a layer. A layer's average thickness is between 0.025 and 0.5 mm. Each layer of resin must be cured using Ultra Violet (UV) light once it has been applied. The three most common types of technology are included in this process: stereo lithography, digital light processing (DLP), and continuous digital light processing (CDLP). Some limitations of vat polymerization are:

1) It's very expensive
2) Longer time process
3) Limited to photo resin materials

Some advantages of the vat photo polymerization process are:

1) Accuracy and surface finish
2) Comparatively faster process
3) Comparatively larger developed part

11.2.7 Sheet lamination

It's the last type of AM process. It binds layers using ultrasonic welding or an adhesive. Sheet lamination can be categorized into seven types:

a) Laminated object manufacturing (LOM)
b) Selective lamination composite object manufacturing (SLCON)
c) Plastic sheet lamination (PSL)
d) Computer aided manufacturing of laminated engineering material (CAM-LEM)
e) Selective deposition lamination (SDL)
f) Composite-based additive manufacturing (CBAM)
g) Ultrasonic additive manufacturing (UAM)

All the sheet lamination processes have the same working principle. In the sheet lamination process, it has been noted that the layer height depends on the sheet thickness. Limited material options are available. The sheet lamination finishes can vary depending on the material, but this process relatively has low cost compared to other process. Its work involves larger working area. It is essentially inter-grades as a hybrid manufacturing system material handling system is very simple. Finally, after the process, it has been observed that this process is rapid but does require post-processing (Sireesha et al., 2018).

11.3 APPLICATION OF ADDITIVE MANUFACTURING

AM has limitless applications. As firms developed for AM and used it as a multiplier to conventional production, new applications have emerged. This has transformed what is imaginable. There are five industries where AM's exceptional capabilities have transformed production. Some important application of AM such as

11.3.1 Medical

In the medical industry, the demand has shifted from patient to patient, and AM promises tailored medical applications (Gaget, 2018; Langau, 2019). Orthopaedic implants, saw guided, heart valves, dental implants, and other medical implants are popular applications that may all be manufactured via AM (Pettersson et al., 2020).

11.3.2 Energy

The parts we're talking about are the ones that go into making energy conversion devices, or systems that convert one type of energy into another. This type of equipment includes electricity generation technology such as wind turbines and solar panels, as well as other conversion devices such as batteries and generators.

11.3.3 Transportation

In many industries, AM (3D printing) is a hot topic and a source of uncertainty. However, from the standpoint of transportation providers, if just one business—or even a small part of several industries—embraces 3D printing, it might have a significant impact on the scale and structure of transportation demand (Tolosa et al., 2010).

11.3.3.1 Automotive sector

Car manufacturers all over the world are using AM and related technologies for their products. Jim Kor, a pioneer of 3D printing in automobile design, is working on a fully electric 3D printed urban city car. By using the AM process, we can improve the automotive sector (Frick, 2014; Jägle et al., 2014).

11.3.3.2 Aerospace/aviation sector

Because the aircraft industry is very much technologically advanced, it has a long history of using AM for the development and production of parts. The benefit of weight reduction is the most important aspect of AM application in aerospace. The reduction in stock and material resource storage space is significant. The cost of upkeep is also vital to consider. It allowed for repeatability and was utilized in a trial-and-error manner to avoid analytical and simulation delays. Essentially, AM is transforming both how and

Table 11.2 Additive manufacture of the aircraft and automobile components

Additive manufactured parts	Materials	Process used	Additive manufactured parts	Materials	Process used
Additive manufactured aerospace parts			Additive manufactured automobile parts		
Jet engine	Ti and Al	Selective laser melting (SLM)/direct metal laser sintering (DMLS)	Pump and valve	Al alloys	SLM and electron beam melting (EBM)
Engine cylinder	Glass-filled nylon	SLS	Engine component	Steel, Ti, or Al alloys	SLM and EBM
Dashboard	Digital ABS	Material jetting	Framework and doors	Al alloys	SLM
Bracket	Castable resin	SLA	Bracket and bumpers	Polymers	SLM
Headlight	Resin	Material jetting and SLA	Headlight	Resin	SLA

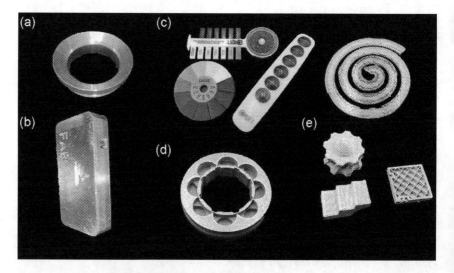

Figure 11.6 The various additive manufacturing–fabricated products in the medical field. (a) Direct energy deposition repairing part, multi-material parts, (b) with sheet lamination, (c) material jetting, (d) binder jetting, and (e) material extrusion of metals and composite parts (Salmi, 2021).

Processing techniques, principles, and applications 199

what we can make. To innovate in a new approach, quality designers and engineers are required. There are even unresolved issues and challenges to overcome, just like prototyping, cost and productivity, complexity, customization, environmental impact, moving first production rate, etc. (Griffiths, 2014; Michelle, 2021). Table 11.2 shows the additive manufacturer's details. Figure 11.6 shows various AM-fabricated products in the field of medical.

11.4 DEVELOPMENT OF ADDITIVE MANUFACTURING

We can't expect those breakthroughs to happen on their own, and we can't expect AM to go away without a lot of effort. We also can't expect manufacturers that are still sceptical of 3D printers to embrace them unless there is some indication that their efforts will pay off. There are three things we need to keep in mind to develop this.

11.4.1 Build process

Consider reorganizing your staff teams so that they can concentrate on the manufacturing process rather than specific items. After arranging the manufacturing plans for various items, we must determine which ones can use 3D printing and form teams that support the required technologies rather than having specific ownership of individual products.

11.4.2 Part validation

Certification of AM parts, particularly for critical applications, is given equal weight. Begin by considering the long-term viability of 3D printing of various elements. If we build them, we can save money and time by using a 3D printer instead of traditional methods. Many industries have developed as a result of this process. Without those standards, proper quality controls have not been a priority, but they are important and will most certainly be implemented soon. Manufacturing can overcome this problem by focusing on the entire triangle—reliability, efficiency, and quality.

11.4.3 Conduct virtual prototype testing

By keeping nonviable designs determined to the drawing board, this testing will help decrease blind alley designs. Digital design files can be evaluated with various materials using virtual prototype test software to see if each concept is practical. The merits and limitations of various types of AM are distinct. It's primarily appropriate for product development, which isn't an easy task. As the pandemic's effects on traditional production continue to spread, the use of AM will become ever more important. As more and more manufacturers experiment with AM and develop their own commercial applications, we're getting closer to achieving economies of scale. This could represent the tipping point for this well-established technology, as well as a significant stride forward toward the fourth industrial revolution.

11.5 SUMMARY AND FUTURE WORK OF ADDITIVE MANUFACTURING

AM 3D printing will revolutionize manufacturing, automobile, airplane, medical, and electronic industries. All the AM processes are well developed and adopted for industries described in this chapter. For selecting a specific object with a specific application, the advantages and limitations of AM are discussed. Food could be printed in the future with personalized nutritional composition based on physiological data. AM, without a doubt, will have an impact on engineering and medical jobs. AM will be increasingly used by engineers working in biomedical, food, automotive, and avionics, as well as civil engineering and industrial design. AM (3D printers) will be able to produce more complex plastics as chemical science advances. AM will almost certainly play a part in the future of any engineering field.

11.6 FUNDING DECLARATION

There was no specific grant for this research from any funding source.

REFERENCES

Campbell, T. A., & Ivanova, O. S. (2013). 3D printing of multifunctional nanocomposites. *Nano Today*, 8(2), 119–120.

Colorado, H. A., Velásquez, E. I. G., & Monteiro, S. N. (2020). Sustainability of additive manufacturing: The circular economy of materials and environmental perspectives. *Journal of Materials Research and Technology*, 9(4), 8221–8234.

Frick, L. (2014). Aluminum-powder DMLS-printed part finishes race first.

Gaget, L. (2018). Medical 3D printing: How 3D printing is saving lives. *Sculpteo*.

Gale, J., & Achuhan, A. (2017). Application of ultrasonic peening during DMLS production of 316L stainless steel and its effect on material behavior. *Rapid Prototyping Journal*, 23(6), 1185–1194.

Gibson, I., Rosen, D. W., & Stucker, B. (2010). *Additive Manufacturing Technologies*. Springer.

Gibson, I., Rosen, D., Stucker, B., & Khorasani, M. (2021). Development of additive manufacturing technology. In *Additive Manufacturing Technologies*, 23–51.

Griffiths, L. (2014). Altair aids RUAG Space with 3D printed satellite support.

Gu, D. (2015). Laser additive manufacturing (AM): Classification, processing philosophy, and metallurgical mechanism. In *Laser Additive Manufacturing of High-Performance Materials*, 8(3), 15–71.

Gu, D. D., Meiners, W., Wissenbach, K., & Poprawe, R. (2012). Laser additive manufacturing of metallic components: Materials, processes and mechanisms. *International Materials Reviews*, 57(3), 133–164.

Hopkinson, N., Hague, R. J. M., & Dickens, P. (2006). *Rapid Manufacturing: An Industrial Revolution for the Digital Age*. Wiley: New York, NY.

Huang, R., Riddle, M., Graziano, D., Warren, J., Das, S., Nimbalkar, S., Cresko, J., & Masanet, E. (2016). Energy and emissions saving potential of additive manufacturing: The case of lightweight aircraft components. *Journal of Cleaner Production*, 135, 1559–1570.

Jägle, E. A., Choi, P.-P., Van Humbeeck, J., & Raabe, D. (2014). Precipitation and austenite reversion behavior of a maraging steel produced by selective laser melting. *Journal of Materials Research*, 29(17), 2072–2079.

Jiao, L., Chua, Z. Y., Moon, S. K., Song, J., Bi, G., & Zheng, H. (2018). Femtosecond laser produced hydrophobic hierarchical structures on additive manufacturing parts. *Nanomaterials*, 8(8).

Langau, L. (2019). For metal additive manufacturing, medical drives demand.

Ma, M., Wang, Z., Wang, D., & Zeng, X. (2013). Control of shape and performance for direct laser fabrication of precision large-scale metal parts with 316L Stainless Steel. *Optics & Laser Technology*, 45, 209–216.

Manfredi, D., Calignano, F., Krishnan, M., Canali, R., Paola, E., Biamino, S., Fino, P. (2014). Additive manufacturing of Al alloys and aluminium matrix composites (AMCs). In *Light Metal Alloys Applications*.

Michelle, J. (2021). Additive Manufacturing in aerospace is growing. *3Dnatives, Business, News, Research*.

Nastac, M., Lucas, R., & Klein, A. (2017). Microstructure and mechanical properties comparison of 316L parts produced by different additive manufacturing processes. In *Proceedings of the 28th Annual International Solid Freeform Fabrication Symposium*, 332–341.

Olsén, J., Shen, Z., Liu, L., Koptyug, A., & Rännar, L. E. (2018). Micro- and macro-structural heterogeneities in 316L stainless steel prepared by electron-beam melting. *Materials Characterization*, 141, 1–7.

Oropeza, D., & Hart, A. J. (2021). A laboratory-scale binder jet additive manufacturing testbed for process exploration and material development. *International Journal of Advanced Manufacturing Technology*, 114(11–12), 3459–3473.

Peng, T., Kellens, K., Tang, R., Chen, C., & Chen, G. (2018). Sustainability of additive manufacturing: An overview on its energy demand and environmental impact. *Additive Manufacturing*, 21, 694–704.

Pettersson, A., Salmi, M., Vallittu, P., Serlo, W., Tuomi, J., & Mäkitie, A. A. (2020). Main clinical use of additive manufacturing (three-dimensional printing) in Finland restricted to the head and neck area in 2016–2017. *Scandinavian Journal of Surgery*, 109, 166–177.

Salmi, M. (2021). Additive manufacturing processes in medical applications. *Materials*, 14(1), 1–16.

Sharma, S. K., Biswas, K., & Dutta Majumdar, J. (2020a). Studies on electron beam surface remelted Inconel 718 superalloy. *Metals and Materials International*.

Sharma, S. K., Biswas, K., & Dutta Majumdar, J. (2020b). Effect of heat input on mechanical and electrochemical properties of electron-beam-welded Inconel 718. *Journal of Materials Engineering and Performance*, 29, 1706–1714.

Sharma, S. K., Biswas, K., Nath, A. K., Manna, I., & Majumdar, J. D. (2020c). Microstructural characterization of laser surface-melted Inconel 718. *Journal of Optics (India)*, 49, 494–509.

Sireesha, M., Lee, J., Kranthi Kiran, A. S., Babu, V. J., Kee, B. B. T., & Ramakrishna, S. (2018). A review on additive manufacturing and its way into the oil and gas industry. *RSC Advances*, 8(40), 22460–22468.

Song, R., Xiang, J., & Hou, D. (2011). Characteristics of mechanical properties and microstructure for 316L austenitic stainless steel. *Journal of Iron and Steel Research, International*, 18(11), 53–59.

Tan, C., Zhou, K., Ma, W., Attard, B., Zhang, P., & Kuang, T. (2018). Selective laser melting of high-performance pure tungsten: Parameter design, densification behavior and mechanical properties. *Science and Technology of Advanced Materials*, 19(1), 370–380.

Tolosa, I., Garciandía, F., Zubiri, F., Zapirain, F., & Esnaola, A. (2010). Study of mechanical properties of AISI 316 stainless steel processed by "selective laser melting", following different manufacturing strategies. *International Journal of Advanced Manufacturing Technology*, 51(5–8), 639–647.

Vafadar, A., Guzzomi, F., Rassau, A., & Hayward, K. (2021). Advances in metal additive manufacturing: A review of common processes, industrial applications, and current challenges. *Applied Sciences*, 11(3).

Vayre, B., Vignat, F., & Villeneuve, F. (2012). Metallic additive manufacturing: State-of-the-art review and prospects. *Mechanics & Industry*, 13(2), 89–96.

Wroe, J. (2015). Introduction to additive manufacturing technology: A guide for designers and engineers. *European Powder Metallurgy Association*, 28–34.

Xie, F., He, X., Cao, S., & Qu, X. (2013). Structural and mechanical characteristics of porous 316L stainless steel fabricated by indirect selective laser sintering. *Journal of Materials Processing Technology*, 213(6), 838–843.

Yusuf, S. M., Chen, Y., Boardman, R., Yang, S., & Gao, N. (2017). Investigation on porosity and microhardness of 316L stainless steel fabricated by selective laser melting. *Metals*, 7(2), 1–12.

Zhong, Y., Rännar, L. E., Liu, L., Koptyug, A., Wikman, S., Olsen, J., Cui, D., & Shen, Z. (2017). Additive manufacturing of 316L stainless steel by electron beam melting for nuclear fusion applications. *Journal of Nuclear Materials*, 486(January), 234–245.

Index

acoustic sensing methods 66
additive manufacturing 21
additive manufacturing techniques 172
advantages and limitations 189
advantages, and limitations of AM 53
advantages of AM over conventional
manufacturing 125
air gap 31
application 35, 54
application of artificial intelligence
(AI) in additive
manufacturing 157
application of 3D printing in aero space
industry 89
application of 3D printing in
automobile industry 92
application of 3D printing in
biomedical domain 90
application of 3D printing in consumer
products 89
application of 3D printing in food
industry 90
applications of additive manufacturing
biomedical domain 122, 161

background: additive
manufacturing 58
bead geometry 64
biomaterials 120
build orientation 30
build time 32

case study 11, 110
ceramics 84
challenges during WAAM process 141

classification of the additive
manufacturing process
118, 191
classifications and fabrication
techniques 45
clinical uses of rapid prototyping 106
closure 14
common defects in titanium alloys 142
complications in conventional
dentistry 102
conclusion 70, 94, 142, 182
conclusion and future scope 37
cost-effective 126
cost estimation 110

defect classification 158
defect detection 66
defect detection using AE 68
dental inserts 100
dentistry 123
deposition depth (H) 10
diagnostic methods 108
directed energy deposition 191

effect of processing parameters 5, 49
efficiency in pre-fabrication 157
electron beam melting (EBM) 48, 152
energy 197
energy consumption 24
extrusion printing 87
extrusion temperature 32

fracture toughness 127
fused deposition modelling 23, 79
future scope 114, 127, 162, 182, 200

203

204 Index

gas metal arc welding 137
gas tungsten arc welding 136

hatch spacing 7
hemorrhages 102
heterogeneous composition 169
homogeneous composition 168

implantation result 112
implant design 109
implant logy 107
implants 122
implant tissue interface 124
infill pattern 28
influence of AM process parameters 4
inspections of powder deposition 62
introduction 21, 43, 57, 77, 100, 118, 150

jet printing 88

laser beam melting 150
laser metal deposition 152
laser power 6
layer height 27
limitation and future scope 14, 93
limitations and challenges 36
limitations of AM in the biomedical domain 126
limited material option 126
low dimensional accuracy 126

material extrusion 177, 192
material jetting 179, 192
materials and technologies used in AM 79
material selection for implant 109
materials in FDM 35
maxillofacial prosthesis 107
mechanical properties 52
mechanical properties of biomedical parts 126
medical 197
melt pool observation 63
metal additive manufacturing techniques 2
microstructure 49, 153
microstructure analysis 139, 140
microstructure LMD 156
microstructure of EBM 154
microstructure of LBM 153
multi-jet printing 87

neurosensory 102

observation 67
operative planning 124
oral medical procedure 107
orthodontics 101, 106,
orthopedics 122

part validation 199
personalized protective equipment 124
pharmaceuticals 123
poly jet 3D printing 81
polymers 79
poor mechanical properties 126
post processing techniques 35
powder bed fusion 46, 179, 192
powder jetting 195
predictive maintenance 160
principle of additive manufacturing 189
printability 157
process parameters 25
product inspection 159

raster angle 26
raster width 25
real-time build control 159
recent advancements 89
recent developments 150
restoration 103
root canal treatment 100

scaling 103
scan orientation 9
selective laser melting (SLM) 47
selective laser sintering (SLS) 47
sensors 161
service-oriented architecture 158
sheet lamination 196
software used 110
stereo lithography 81
strategies for quality improvement 142
summary and future scope 54, 81
surgical implants 161
surgical tools 124

temperature monitoring 65
tensile and compressive strength 127
tensile properties analysis 139, 140
thermal sensing methods 65
tissue engineering 122
tissue regeneration 122
titanium alloys fabricated using CMT 140
titanium alloys fabricated using GTAW 139

Index 205

topology 162
traditional methods in dental treatment 100
transportation 197
types of defects 158

use of rapid prototyping in dentistry 106
uses of stereo lithographic models 104

VAT photo polymerization 173, 195
virtual prototype 199
vision sensing methods 61

waste reduction 160
wire arc additive manufacturing 135, 138
working concept 105